航空发动机专业英语

主 编 王 姝 顾鸣惊
参 编 王洪明

华中科技大学出版社
中国·武汉

内 容 简 介

本书由三大模块组成：模块一为飞机动力装置概述，内容涉及航空发动机的分类、航空发动机的发展历史、全球具有代表性的民用航空发动机及燃气轮机发动机的主要参数及其基本要求；模块二为航空发动机结构，内容涉及进气道、压气机、燃烧室、涡轮、喷管等航空发动机主要结构；模块三为航空发动机系统，内容涉及航空发动机燃油系统、航空发动机滑油和冷却系统、航空发动机启动和点火系统、航空发动机防火系统、航空发动机指示系统和防冰系统等航空发动机的主要系统。通过学习本书，学生可以掌握主要的航空发动机英语专业词汇等，还能对航空发动机原理与构造有初步的了解。

本书可以作为高职类航空院校的航空发动机制造技术、航空发动机装配与调试技术、航空发动机维修技术等专业的专业英语教材，也可作为飞机维修、飞行器数字化制造技术、通用航空等专业的拓展英语资料。

图书在版编目(CIP)数据

航空发动机专业英语 / 王姝，顾鸣惊主编. -- 武汉：华中科技大学出版社，2025.1. -- ISBN 978-7-5772-1579-2

Ⅰ. V23

中国国家版本馆 CIP 数据核字第 2025K9N591 号

航空发动机专业英语　　　　　　　　　　　王　姝　顾鸣惊　主编
Hangkong Fadongji Zhuanye Yingyu

策划编辑：王　勇
责任编辑：姚同梅
封面设计：廖亚萍
责任监印：朱　玢

出版发行：华中科技大学出版社(中国·武汉)	电话：(027)81321913	
武汉市东湖新技术开发区华工科技园	邮编：430223	

录　　排：武汉三月禾文化传播有限公司
印　　刷：武汉市洪林印务有限公司
开　　本：787mm×1092mm　1/16
印　　张：9
字　　数：236 千字
版　　次：2025 年 1 月第 1 版第 1 次印刷
定　　价：39.80 元

本书若有印装质量问题，请向出版社营销中心调换
全国免费服务热线：400-6679-118　竭诚为您服务
版权所有　侵权必究

前　　言

航空发动机是一种高度复杂和精密的热力机械,作为航空器的"心脏",不仅为航空器提供飞行所需动力,也为促进航空事业发展提供了重要推动力。航空发动机直接影响飞机的性能、可靠性及经济性,航空发动机的发展水平是一个国家科技、工业和国防实力的重要体现。随着全球航空工业的快速发展,航空发动机的技术复杂性和精度要求日益提升,业内对航空发动机制造和维修等从业人员的英语阅读水平、翻译水平等也提出了更高的要求。

虽然相关院校都意识到航空发动机制造、维修等专业的专业英语的重要性,也认识到合适的专业英语教材是提升学生英语水平和能力的重要基础,但遗憾的是,目前市场上合适的此类专业英语教材种类并不多。为了培养具有国际视野、能够与国际同行交流合作的航空发动机专业人才,我们编写了本书。本书旨在提高航空发动机相关专业学生的英语应用能力,特别是专业英语阅读和写作技能。我们深知,在经济全球化背景下,掌握专业英语不仅是与国际同行沟通的需要,更是推动航空发动机技术创新的必要条件。

航空发动机的结构比较复杂,主要由进气装置、压气机、燃烧室、涡轮和排气装置五大部分组成。其中,压气机、燃烧室和涡轮是三大核心部件,它们共同组成航空发动机的核心。本书分为三大模块,每个模块均围绕航空发动机的核心部件展开。全书共计12课,旨在通过详细而全面的英文介绍,帮助学生掌握航空发动机专业的主要英语专业词汇,并初步理解航空发动机的原理与构造。希望通过本教材的学习,学生能够掌握航空发动机专业英语的基本知识和技能,提高在国际航空工业中的竞争力。

本书有以下方面特色:

(1) 基于航空发动机制造、维修职业能力需求开发学习内容。

我们从深入了解航空发动机制造、维修等岗位能力需求出发,围绕职业能力培养,紧扣航空发动机专业的核心知识点,选取了一批最具代表性的英文文献和资料,力求使学生在学习过程中既能够掌握专业英语词汇和表达方式,又能够了解国际前沿的航空发动机技术动态。

(2) 注重内容的系统性与创新性。

本书在内容安排上力求系统全面,覆盖航空发动机的各个方面。同时,我们也注重创新性的培养,鼓励学生在学习中发挥主动性和创造性,培养批判性思维能力和创新能力。

（3）依托航空发动机结构安排学习情境。

本书依托航空发动机主要结构来编排学习情境,通过学习这些情境英语术语和高频句型,学生将循序渐进地掌握相关学习内容。

（4）配以实物图片,讲解直观生动。

本书通过将专业讲解与实物插图相结合,图文并茂地将航空发动机主要结构和系统展现出来,使学生能比较容易把抽象的英文术语与实物相联系,从而使学习更加高效。

（5）教学资源丰富。

为满足教学要求和方便学生学习,激发学生学习潜能,培养学生听、说的能力,便于学生加强对教材内容的理解和记忆,本书以二维码形式提供了课文音频和课文翻译。

本书可以作为高职类航空院校的航空发动机制造技术、航空发动机装配与调试技术、航空发动机维修技术等专业的专业英语教材,也可作为飞机维修、飞行器数字化制造技术、通用航空等专业的拓展英语资料。

本书由王姝、顾鸣惊主编,王洪明参编。其中第1—6课由王姝编写,第8—11课由顾鸣惊编写,第7、12课由王洪明编写。王姝负责全书统稿工作。本书获得了民航发展基金教育人才类项目资金支持,在此表示衷心感谢。同时,也向对本教材的编写给予指导、支持、帮助的领导,飞机制造工程教研室的同事,以及企业专家致以诚挚的谢意。

由于编者实践经验有限,书中难免存在疏漏与不足,敬请广大读者和专家批评指正,编者会在本书使用过程中仔细检查,并一一校正。同时,也期待后续与广大师生共同探讨和完善本书的内容,为培养更多的航空发动机专业人才贡献力量。

编　者

2024年11月

Contents

Module One Overview of Aircraft Powerplants 1
   Lesson 1   Classifications of Aero Engines 1
   Lesson 2   The Development of Aero Engines 12

Module Two Structure of Aero Engine 22
   Lesson 3   Air Inlet Duct 22
   Lesson 4   Compressors 30
   Lesson 5   Combustion Section 43
   Lesson 6   Turbine Section 52
   Lesson 7   Exhaust Section 59

Module Three Engine System 69
   Lesson 8   Fuel System 69
   Lesson 9   Lubrication and Cooling Systems 78
   Lesson 10   Starting and Ignition Systems 90
   Lesson 11   Engine Fire-protection System 105
   Lesson 12   Engine Indication System and Ice Protection System 114

Glossary of Commonly-used Terms 122

课后习题答案 132

参考文献 138

Module One Overview of Aircraft Powerplants

 ## Lesson 1 Classifications of Aero Engines

课文音频

The first man-carrying flight was made in a hot air balloon swept along by air currents and without means for the pilot to control the direction of flight. Aircraft had little practical utility until the development of engine-driven propellers. This development of the powerplant has made aviation a vital factor in today's economic world. The powerplant includes the complete installation of the aero engine (also know as the aircraft engine), propeller, and all accessories needed for its proper function.

After more than a hundred years of development, the aero engine has developed into a mature product with high reliability. The aero engine used by humans is mainly divided into two categories: the reciprocating engine and the gas turbine engine, as shown in Figure 1.1. All powered aircraft are driven by some form of heat engine. Chemical energy stored in the fuel is released as heat energy that causes air to expand. The expansion of air is what performs useful work, driving either a piston or a turbine. There are two basic types of heat engines: external-combustion and internal-combustion. External-combustion engines are most familiar to us as steam engines. Energy released in coal- or gas-fired furnaces or in nuclear reactors is transferred into water, changing it into steam that expands and drives either a piston or a turbine. Internal-combustion engine is a form of heat engine in which the fuel and air mixture is burned inside the engine. Internal-combustion engines release energy from fuel directly inside an engine to heat and expand the air. The expanding air can drive reciprocating pistons or spin turbines. Most of the current aero engines are of the internal combustion type because the combustion process takes place inside the engine.

Reciprocating engine

A reciprocating engine is a mechanical device that mixes fuel and air in a closed container, releases heat energy through combustion, and then the hot gas expands to

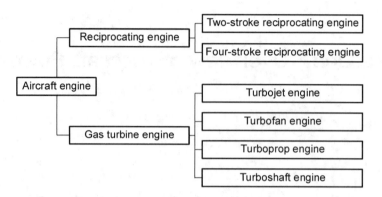

Figure 1.1 Categories of aero engines

do work on the connecting rod. A propeller is installed on a reciprocating engine to generate pushing (pulling) force. So, as the power device of an aircraft, reciprocating engines and propellers are inseparable. Reciprocating engines can be divided into two categories based on the number of piston strokes required to complete a job: two-stroke reciprocating engines and four-stroke reciprocating engines. The stroke is the distance the piston moves from one end of the cylinder to the other, specifically from top dead center (TDC) to bottom dead center (BDC), or vice versa, as shown in Figure 1.2. A two-stroke reciprocating engine requires the piston to reciprocate two strokes to complete a working cycle, while a four-stroke reciprocating engine requires the piston to reciprocate four strokes to complete a working cycle.

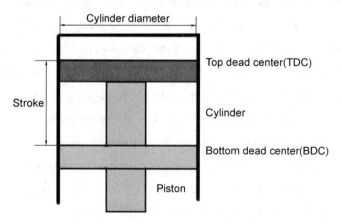

Figure 1.2 Stroke

The major components of a reciprocating engine are the crankcase, cylinder, piston, connecting rod, valve, valve-operating mechanism, and crankshaft. In the head of each cylinder are the valves and spark plugs. One of the valves is in a passage leading from the induction system, the other is in a passage leading to the exhaust system. Inside each cylinder is a movable piston connected to a crankshaft by a connecting rod. Figure 1.3 illustrates the basic parts of a reciprocating engine.

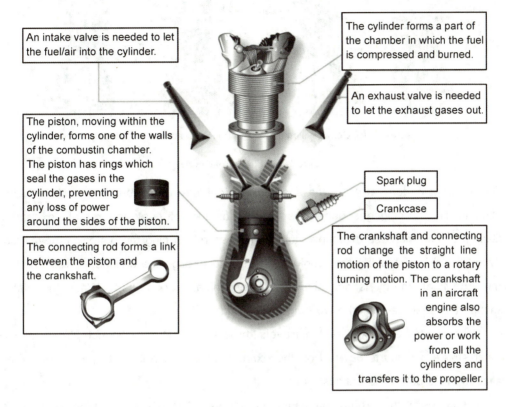

Figure 1.3 Basic parts of a reciprocating engine

Modern aviation reciprocating engines belong to four-stroke engines. A four-stroke reciprocating engine completes four processes within four strokes: intake, compression, power, and exhaust (Figure 1.4). Two complete revolutions of the crankshaft (720°) are required for the four strokes' thus, ignition occurs once in each cylinder for every two revolutions of the crankshaft.

Intake Stroke: During the intake stroke, the piston is pulled downward in the cylinder by the rotation of the crankshaft. This reduces the pressure in the cylinder and causes air under atmospheric pressure to flow through the carburetor, which meters the correct amount of fuel. The fuel/air mixture passes through the intake pipes and intake valves into the cylinders.

Compression Stroke: After the intake valve is closed, the continued upward travel of the piston compresses the fuel/air mixture to obtain the desired burning and expansion characteristics.

Power Stroke: When the piston approaches the BDC of the compression stroke, an

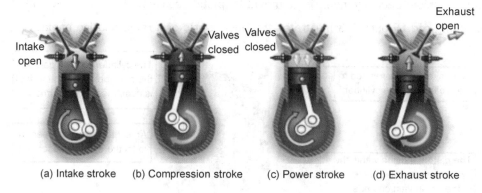

(a) Intake stroke　　(b) Compression stroke　　(c) Power stroke　　(d) Exhaust stroke

Figure 1.4　Four-stroke cycle

electric spark is generated by the spark plug, igniting the combustible mixture in the cylinder. The burning gas generates a large amount of heat and rapidly expands, causing a rapid increase in temperature and pressure, forcing the piston to move from top to bottom and pushing the crankshaft to rotate through the connecting rod. At this time, the heat energy is converted into mechanical work output. When the piston reaches the dead center, the working stroke ends.

Exhaust Stroke: As the piston travels through the BDC at the completion of the power stroke and starts upward on the exhaust stroke, it begins to push the burned exhaust gases out of the exhaust port.

The speed of the exhaust gases leaving the cylinder creates a low pressure in the cylinder. This low or reduced pressure speeds the flow of the fresh fuel/air charge into the cylinder as the intake valve is beginning to open.

Gas turbine engine

The main feature of a gas turbine engine is that it relies on high-temperature and high-speed gas to drive the turbine, which drives the compressor to compress the air. In a reciprocating engine, the functions of intake, compression, combustion, and exhaust all take place in the same combustion chamber. Consequently, each must have exclusive occupancy of the chamber during its respective part of the combustion cycle. A significant feature of the gas turbine engine is that separate sections are devoted to each function, and all functions are performed simultaneously without interruption.

A typical gas turbine engine consists of the air inlet, compressor section, combustion section, turbine section, exhaust section, accessory section, and the systems necessary for starting, lubrication, fuel supply, and auxiliary purposes, such as anti-icing, cooling, and pressurization.

The major components of all gas turbine engines are basically the same; however, the nomenclature of the component parts of various engines currently in use varies slightly due to the difference in each manufacturer's terminology. These differences are reflected in the applicable maintenance manuals. One of the greatest single factors influencing the construction features of any gas turbine engine is the type of compressor or compressors for which the engine is designed. Gas turbine engines are the power plants for all the flying aircraft/helicopters (and in this case may be denoted as aero engines or aero derivative gas turbines) and sources for power in miscellaneous industrial applications in automobiles, tanks, marine vessels, and electric power generation devices. The classification of gas turbine engines are shown in Figure 1.5. At present, there are mainly four basic types of aviation gas turbine engines: turbojet engines, turbofan engines, turboprop engines, and turboshaft engines.

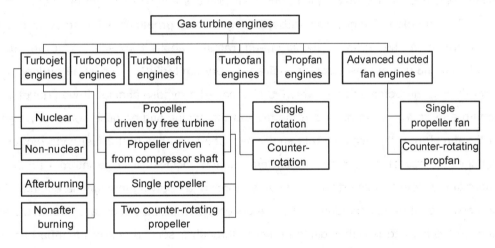

Figure 1.5 Classification of gas turbine engines

(1) Turbojet Engine: A typical turbojet engine consists of six basic components (Figure 1.6): air inlet duct, compressor, combustor, diffuser, turbine, and exhaust section(tail pipe and jet nozzle). The compressor takes air in through the aircraft air inlet duct and increases its pressure. A turbojet engine then sends this compressed air through a diffuser into a burner, or combustor. Fuel is then added and burned, and the resulting expanded and heated air flows through a turbine that extracts some of the energy to drive the compressor. A high-velocity stream of exhaust gases leaves the engine through a specially shaped exhaust nozzle where it produces thrust. The amount of thrust is proportional to the increase in the momentum of air that passes through the engine.

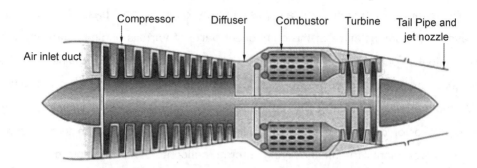

Figure 1.6 Basic components of a turbojet engine

The turbojet engine has greatly reduced the expense of air travel and improved aircraft safety. The turbojet engine also allows the aircraft to have faster speeds, even supersonic speeds. It has a much higher thrust per unit weight ratio than the piston-driven engine, which leads directly to longer ranges, higher payloads, and lower maintenance costs. Military fighters and fast business jets use turbojet engines.

(2) Turbofan Engine: The turbojet engine has problems with noise and fuel consumption in the airlines' flying speed range (about 8 Mach). Due to these problems, the use of pure turbojet engines is very limited. So, almost all airliner type aircraft use a turbofan engine, which is also called a bypass engine. It was developed to turn a large fan or set of fans at the front of the engine and produce about 80 percent of the thrust from the engine. This engine is quieter and has better fuel consumption in the speed range of the airliner. A turbofan engine includes a large internal propeller (normally denoted a ducted fan) and two streams of air flowing through the engine. The primary stream travels through all the components, while the secondary stream passes through the fan and is either ducted outside through a second nozzle identified as the cold nozzle or may mix with the hot gases leaving the turbine and both are expelled from a single nozzle (Figure 1.7). Turbojet engines produce thrust by accelerating a smaller mass of air by a larger amount, and turboprop engines accelerate a larger mass by a smaller amount. A turbofan engine produces thrust similar to that produced by a combination of turbojet and turboprop engines, which has a set of lengthened blades in the first stage or stages of the low-pressure compressor (Figure 1.8). Air flows through the fan section of the engine bypasses, or flows around the outside of the core engine. The amount of thrust produced by the fan varies between 30% and 75% of the total thrust, depending upon the bypass ratio. The bypass ratio, which is the ratio of the mass of air moved by the fan to the mass of air moved by the core engine, varies from less than 1 : 1 for low-bypass engines, up to about 8 : 1 for some high-

bypass engines. Some low-bypass turbofan engines have long annular fan-discharge ducts along the full length of the engine. The fan-discharge air and the exhaust from the core engine are discharged separately in some long-duct engines, and in others, the two discharges are mixed before leaving the engine.

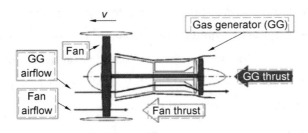

Figure 1.7 Turbofan layout

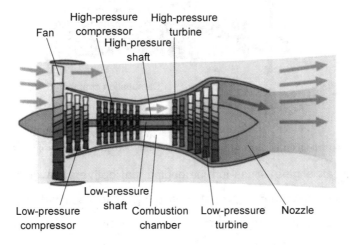

Figure 1.8 Basic components of a turbofan engine

(3) Turboprop engine: Turboprop engines are used for many single channel aircraft, dual channel aircraft, and commuter aircraft. The turbofan engine produces thrust directly; the turboprop engine produces thrust indirectly — the compressor and turbine assembly furnishes torque to a propeller, producing the major portion of the propulsive force that drives the aircraft. Turboprop engines combine the best features of turbojet and piston engines. The former is more efficient at high speeds and high altitudes, while the latter is more efficient at speeds under 400 — 450 mph and altitudes below 30,000 ft. Consequently, commuter aircraft and military transports tend to feature turboprop engines.

A turboprop engine differs from a turbojet engine in that the design is optimized to produce rotating shaft power to drive a propeller, instead of thrust from the exhaust gas. The turboprop engine uses a gas turbine to turn a large propeller. The shaft that connects the propeller to the turbine is also linked to a gearbox that controls the propeller's speed. The propeller is most efficient and quiet, when the tips are spinning at a speed just below supersonic speed. Moreover, no propeller is capable of withstanding the forces generated when it turns at the same speed of the turbine. Turboprop engines may be further classified into two groups, depending on the turbine driving the propeller. In the first group, the propeller is driven by the same gas turbine driving the compressor. In the second group, additional turbine (normally denoted as free power turbine) turns the propeller. Fig 1.9 illustrates a turboprop engine with a single turbine driving the compressor and propeller.

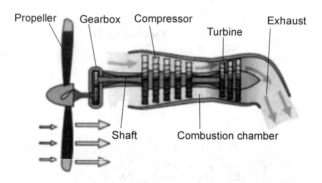

Figure 1.9 A turboprop engine with a turbine driving both the compressor and the propeller

(4) Turboshaft engine: A gas-turbine engine that delivers power through a shaft to operate something other than a propeller is referred to as a turboshaft engine. The output shaft may be coupled directly to the engine turbine, or the shaft may be driven by a turbine of its own (free turbine) located in the exhaust stream. The free turbine rotates independently. This principle is used extensively in current production of turboshaft engines. The turboshaft engine's output is measured in horsepower instead of thrust because the power output component is a turning shaft.

Turboshaft engines are ideally suited for powering helicopters, because they operate most efficiently at the constant RPM required by a helicopter. Most turboshaft engines drive their output shaft with a multi-stage free turbine that extracts as much energy as possible from the exhaust gases, as shown in Figure 1.10. The turbines drive shafts which are used to drive helicopter rotors, generators, or pumps. The general layout of a turboshaft is similar to that of a turboprop, the main difference being that the latter produces some residual propulsion thrust to supplement the thrust produced by the shaft-driven propeller. Residual thrust on a turboshaft engine is avoided by further expansion in the turbine system and/or truncating and turning the exhaust through 90°. Another difference is that for a turboshaft the main gearbox is part of the helicopter while for a turboprop the gearbox is a part of the engine. Virtually all turboshafts have a "free" power turbine, although this is also generally true for modern turboprop engines.

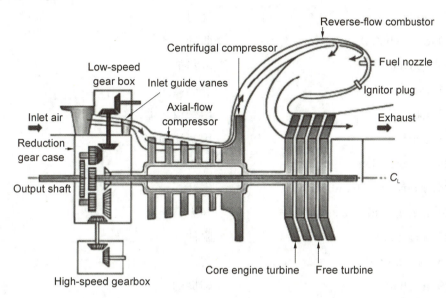

Figure 1.10　A turboshaft engine that uses a multi-stage free turbine to drive the output shaft

 Words & Expressions

1. pilot　　　　　　　　　　　飞行员
2. propeller　　　　　　　　　螺旋桨
3. powerplant　　　　　　　　动力装置
4. aircraft　　　　　　　　　　飞机,航空器
5. aircraft engine　　　　　　 航空发动机
6. reciprocating engine　　　活塞发动机
7. turbine　　　　　　　　　　涡轮
8. gas turbine engine　　　　燃气涡轮发动机
9. combustion　　　　　　　　燃烧
10. external-combustion　　　外燃机的,外燃的
11. internal-combustion　　　 内燃机的,内燃的
12. turbojet engine　　　　　 涡轮喷气发动机

课文翻译

13.	turbofan engine	涡轮风扇发动机
14.	turboprop engine	涡轮螺旋桨发动机
15.	turboshaft engine	涡轮轴发动机
16.	crankcase	曲轴箱，机匣
17.	rod	连杆，手柄
18.	valve	气门
19.	crankshaft	曲轴
20.	coolant	冷却剂
21.	radiator	散热器
22.	tube	用管道运输
23.	bolt	用螺栓固定
24.	compression	压缩
25.	chamber [ˈtʃeɪmbə(r)]	室，腔
26.	horsepower	马力
27.	accessory	附件
28.	lubrication	润滑
29.	auxiliary	辅助的
30.	maintenance	维修，保养
31.	diffuser	扩散器
32.	velocity	速度
33.	thrust	推力
34.	supersonic	超声速的
35.	payload	载荷
36.	jet	喷气式飞机
37.	bypass ratio	涵道比
38.	assembly	装配组件
39.	propulsive	推进的
40.	propeller governor coordination	螺旋桨调速器协作
41.	mph	每小时英里数
42.	ft	英尺
43.	gearbox spin	齿轮箱

Exercise

Ⅰ. **Fill in the following blanks according to the text.**

1. Aircraft had little practical utility until the development of engine-driven _____.

2. The _____ includes the complete installation of an aero engine, propeller, and all accessories needed for its proper function.

3. _____ Stroke: After the intake valve is closed, the continued upward travel of the piston compresses the fuel/air mixture to obtain the desired burning and expansion characteristics.

4. A turbojet engine then sends this compressed air through a _____ into a burner, or combustor. Fuel is then added and burned, and the resulting expanded and heated air flows through a turbine that extracts some of the energy to drive the compressor.

5. The _____, which is the ratio of the mass of air moved by the fan to the mass of air moved by the core engine, varies from less than 1∶1 for low-bypass engines, up to about 8∶1 for some high-bypass engines.

Ⅱ. **Translate the following sentences into Chinese.**

1. Energy released in coal- or gas-fired furnaces or in nuclear reactors is transferred into water, changing it into steam that expands and drives either a piston or a turbine.

2. A reciprocating engine is a mechanical device that mixes fuel and air in a closed container, releases heat energy through combustion, and then the hot gas expands to do work on the connecting rod.

3. The major components of a reciprocating engine are the crankcase, cylinder, piston, connecting rod, valve, valve-operating mechanism, and crankshaft. In the head of each cylinder are the valves and spark plugs.

4. Modern aviation reciprocating engines belong to four-stroke engines. A four-stroke reciprocating engine completes four processes within four strokes: intake, compression, power, and exhaust.

5. At present, there are mainly four basic types of aviation gas turbine engines: turbojet engines, turbofan engines, turboprop engines, and turboshaft engines.

6. Most turboshaft engines drive their output shaft with a multi-stage free turbine that extracts as much energy as possible from the exhaust gases.

Lesson 2　The Development of Aero Engines

2.1　History of Aero Engines

The development of aero engines has played a crucial role in the advancement of aviation technology. From the early days of piston engines to the modern era of jet propulsion, aircraft engines have undergone significant transformations.

For more than a hundred years since the birth of aero engines, two main stages have passed through. The first 40 years (1903－1945) were the ruling period for piston engines. The last 80-odd years (1939－now) were the era of jet engines. During this period, gas turbine engines were widely used in aviation, and turbojet engines and turbofan engines with direct thrust generation were developed successively.

The first aero engine in human history was the 12 horsepower 4-cylinder gasoline engine of the Wright brothers in 1903 (Figure 2.1), which used aluminum cast cylinder blocks to solve the problem of reducing weight while maintaining power. At that time, simple piston engines powered by internal combustion were utilized to provide thrust for aircraft. These early engines were often unreliable and had limited power output, restricting the range and capabilities of early aircraft.

During World War Ⅰ, there was a rapid advancement in aero engine technology driven by the demands of military conflict. Engineers developed more powerful and efficient engines, such as the rotary engines used in planes like the Sopwith Camel. The first rotary engine was a 7-cylinder 50 hp (37 kW) unit (Figure 2.2). Following the end of World War Ⅰ, the interwar period saw continued improvements in aero engine design. Radial engines became popular due to their reliability and performance. In the lead-up to World War Ⅱ, advancements in supercharging and fuel injection led to even more powerful engines.

The introduction of jet engines in the mid-20th century revolutionized aviation. Jet engines offer higher speeds, greater reliability, and improved altitude performance compared to traditional piston engines. The development of turbojet and later turbofan engines further enhanced the capabilities of aircraft. The turbojet engine was invented by Frank Whittle in the 1930s and first flew in 1941. This engine works by compressing air into the combustion chamber where it is mixed with fuel and ignited. The resulting hot gases expand and are then exhausted through a turbine, which spins the compressor and generates thrust.

Figure 2.1　The first wright engine　　　　Figure 2.2　The first rotary engine

In the post-World War II era, aero engine development continued to accelerate with the introduction of new materials and technologies. Engines became more reliable, efficient, and durable, thanks to advances in metallurgy, manufacturing processes, and design techniques.

Today, aero engines are extremely complex machines that combine advanced technologies with high-performance materials to achieve maximum efficiency and reliability. Engines are designed to operate at high temperatures and pressures while providing exceptional thrust-to-weight ratios and fuel efficiency. In recent decades, aero engine technology has focused on increasing fuel efficiency, reducing emissions, and enhancing performance. Innovations such as high-bypass turbofans, composite materials, and digital controls have transformed the aviation industry. Ongoing research aims to develop sustainable alternative fuels and hybrid-electric propulsion systems for future aircraft.

From the humble gasoline engines of the early 20th century to the advanced turbofan and jet engines of today, humans have come a long way in our pursuit of flight. Aero engines are poised for further advancement in the coming decades. Technologies such as electric propulsion, fuel cells, and alternative fuels hold promise for future aero engines, while specialists will continue to address safety and efficiency concerns as they seek to further push the boundaries of aviation.

2.2 Global civil aero engines

Civil aero engines are critical components of aircraft that provide the power necessary for flight. Today, there are numerous manufacturers of civil aero engines, each with their own unique designs and technologies. Some of the most prominent names in the industry include General Electric (GE) Aviation, Pratt & Whitney (P&W), Rolls-Royce, and CFM International. Each manufacturer has its own unique lineup of engines, each with its own characteristics and applications.

GE Aviation is a subsidiary of General Electric and one of the leading suppliers of civil aviation engines. Their offerings include the CF34, GE90, and the new LEAP-X series. Their flagship engine, the GEnx, powers some of the most popular wide-body jets, such as the Boeing 777X. The GE90 is a large-capacity engine from GE Aviation that provides power for long-range flights on Boeing 777 and 787 aircraft.

P&W is a subsidiary of United Technologies Corporation and is one of the leading manufacturers of commercial aero engines. Their products include JT series and PW series. Their PW1000G family of engines is used on some of the most fuel-efficient narrow-body jets, such as the Airbus A320neo.

Rolls-Royce is another major player in the aero engine market, with a wide range of engines available for different types of aircraft. Their Trent family of engines is used on many long-haul narrow-body jets, such as the Boeing 787 and Airbus A320 family.

CFM International is a joint venture between Snecma (Safran Group) and General Electric, and is the leading supplier of engines for the new generation of single-aisle jets, such as the Boeing 737 MAX and Airbus A320neo families. Their LEAP-1A and LEAP-1B engines are some of the most fuel-efficient and environmentally friendly engines available today.

Next, the paragraph mainly introduces commonly used aviation engines in civil aviation, such as CFM56, V2500, and GE90.

CFM56 turbofan engine

CFM56 is a high-bypass ratio, dual rotor, and axial flow turbofan engine commonly mounted on the Boeing 737 aircraft, as shown in Figure 2.3, with a thrust range of 8900 to 10700 daN (1 daN = 10 N).

The engine consists of a 4-stage integral fan, a low-pressure compressor (including 1-stage fan, 1-stage fan outlet guide vanes, 3-stage turbocharger rotor, 4-stage turbocharger stator, 12 adjustable air release valves evenly distributed along the circumference at the outlet for low-power state to put some air into the fan channel), 9-stage high-pressure compressor (including the 1st stage adjustable inlet guide vanes, the 3rd stage adjustable stator blades, the 5th stage

Figure 2.3 CFM56-7 turbofan engine

fixed stator blades), circular combustion chamber. The low-pressure compressor is driven by a 4-stage low-pressure turbine, the 9-stage high-pressure compressor is driven by the 1st stage high-pressure turbine.

The CFM56-7 adopts 24 wide chord fan blades and is designed with a new booster stage and a dual ring combustion chamber. Compared with CFM56-3, noise and pollution are significantly reduced, maintenance costs are reduced by about 15%, and thrust is between 8,684 daN and 11,730 daN.

V2500 turbofan engine

The V2500 turbofan engine is a dual rotor, axial flow, high-bypass turbofan engine with a thrust of 25,000 (11,100 daN) levels.

The low-pressure rotor of the engine includes a 1-stage fan, a 4-stage low-pressure compressor, and a 5-stage low-pressure turbine. The high-pressure rotor includes a 10-stage high-pressure compressor, a 2-stage high-pressure turbine, and a circular combustion chamber. The structural diagram of the V2500 turbojet engine is shown in Figure 2.4.

The engine adopts advanced devices such as wide chord fans, floating wall combustion chambers, single crystal turbine blades and powder metallurgy turbine discs, active control systems of blade tip clearance, and full digital electronic control systems, which enhance the engine's competitiveness. In the model number, "V" represents a collaborative development among five countries, and "2500" represents a thrust level in units of 10 lb (1 lb = 0.454 kg).

GE90 turbofan engine

The GE90 turbofan engine is an engine assembled on the Boeing 777 aircraft, as shown in Figure 2.5. When obtaining the airworthiness certificate, the GE90 had a thrust of 37,675 daN, but when installed on the Boeing 777 aircraft, it operated at reduced power (reduced by 10%), which meant it was put into initial service with a thrust of 34,250 daN. In addition to providing maximum thrust, the GE90 reduces fuel consumption by 9% and reduces nitrogen oxide emissions by 33% compared to existing high-bypass engines.

The engine adopts a single-stage axial flow fan, a 3-stage axial flow low-pressure compressor, a 10-stage axial flow high-pressure compressor, an annular combustion chamber, a 2-stage axial flow high-pressure turbine, and a 6-stage large-diameter axial flow turbine. The control system used is a fully functional digital electronic control system, which controls fuel consumption, turbine radial clearance, and compressor stator blade installation angle.

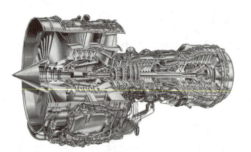

Figure 2.4　V2500 turbofan engine

Figure 2.5　GE90 turbofan engine

LEAP turbofan engine

At present, LEAP turbofan engines have been successfully applied in multiple narrow body aircraft, such as the Boeing 737 MAX, Airbus A320neo, and C919. As the successor of the CFM56 engine, the LEAP turbofan engine has undergone comprehensive upgrades and optimizations in design, aiming to provide more efficient and environmentally friendly power support for narrow body aircraft.

The LEAP turbofan engine features a new-generation compressor design, enhancing efficiency and reliability through optimized blade shapes and fewer stages. Additionally, the LEAP engine incorporates advanced combustion chamber technology, lowering fuel consumption and pollution emissions. Furthermore, the engine utilizes cutting-edge turbine design and material technology, boosting turbine efficiency and durability. In performance terms, the LEAP turbofan engine outperforms the previous generation CFM56 by reducing fuel consumption by approximately 15%, along with decreased carbon dioxide emissions and noise pollution. Consequently, aircraft equipped with LEAP turbofan engines operate more economically and environmentally, delivering significant economic and societal benefits to airlines.

Beyond their performance benefits, LEAP turbofan engines also excel in reliability. Equipped with advanced fault diagnosis and health management technology, these engines can monitor their operational status in real-time, promptly identify and address potential issues. This exceptional reliability and safety have earned LEAP turbofan engines a place as a preferred powerplant for airlines.

2.3 Main parameters of gas turbine engines and their basic requirements

Aero engines convert various forms of energy into mechanical power that propels the aircraft into the air and sustains its flight. Understanding the main parameters is crucial for their design, manufacture, maintenance, and operation.

2.3.1 Main parameters of gas turbine engines

Thrust

Thrust is the key factor determining an aero engine's capacity to propel an aircraft forward. It represents the force produced by the engine, directed in the direction of the aircraft's motion. The higher the thrust, the more rapidly an aircraft can accelerate and sustain its fast speed. However, the magnitude of thrust alone cannot define an engine's performance, as it does not reflect the engine's size, weight, or the fuel efficiency required to generate such thrust. Consequently, parameters like unit thrust and thrust-to-weight ratio have been introduced. The ratio of the thrust of an engine to the mass flow of air flowing through the engine is called the unit thrust or specific thrust. The ratio of the thrust of an engine to the mass of the engine is called the thrust-to-weight ratio.

Under specified conditions, a higher unit thrust corresponds to a greater thrust output from the engine. A higher thrust-to-weight ratio indicates that, with a constant thrust, the engine weighs less; conversely, with a constant engine weight, the thrust is greater. Therefore, a higher thrust-to-weight ratio is preferable.

Rate of fuel consumption

Fuel consumption is the quantity of fuel that enters the combustion chamber per unit time. For a given engine, fuel consumption serves as a critical monitoring parameter because it increases with declining performance and the onset of malfunctions. Consequently, changes in fuel consumption and exhaust temperature can be used to discern engine performance degradation and malfunctions.

The fuel consumption rate denotes the mass of fuel consumed to generate one unit of thrust over the course of one hour. The factors influencing the fuel consumption rate include the unit thrust and the fuel-to-air ratio. With the unit thrust held constant, a higher fuel-to-air ratio results in a higher fuel consumption rate; conversely, a higher unit thrust results in a lower fuel consumption rate when the fuel-to-air ratio remains constant.

The factors influencing the fuel consumption rate include the unit thrust and the temperature differential between the combustion chamber's inlet and outlet. With the temperature differential between the combustion chamber's inlet and outlet held constant, a higher unit thrust leads to a lower fuel consumption rate. Conversely, when the unit thrust remains constant, a larger temperature differential between the combustion chamber's inlet and outlet results in a higher fuel consumption rate.

Total gas temperature before turbine

The total gas temperature before turbine is the most significant and pivotal parameter in a gas turbine engine, albeit being a constrained one. It serves as an indicator of the engine's performance level and is utilized as a monitoring parameter for maintenance purposes. Whenever the total gas temperature before turbine surpasses a specific threshold, maintenance is deemed necessary.

2.3.2 Basic requirements for aero engines

The fundamental requirements of civil aviation for aero engines primarily encompass the following aspects:

High thrust-to-weight ratio and small unit frontal area

The size of the engine's frontal area, under specific flight conditions, determines the magnitude of external drag experienced by the engine nacelles, thereby directly influencing the effective thrust generated. Minimizing the unit frontal area can enhance the efficient utilization of thrust. Elevating the thrust-to-weight ratio necessitates a higher utilization efficiency of the thermal cycle, along with a reduction in the engine's dimensions and mass. This necessitates the utilization of materials with high specific strength and the adoption of reasonable design techniques to the fullest extent possible.

Low unit fuel consumption rate

Fuel consumption constitutes a significant proportion of the direct operating costs of flights. Consequently, there is an increasing demand for improved fuel consumption efficiency in engines, necessitating the continuous reduction of fuel consumption rates. For civil aviation aircraft, enhancing fuel economy not only extends flight range and endurance but also enables an increase in payload capacity within a given range. Therefore, reducing the unit fuel consumption rate remains a crucial aspect of current engine development efforts in the realm of civil aviation.

Stable operation and high reliability

The stability of engine operation is evaluated from a thermodynamic and aerodynamic perspective, taking into account factors such as engine speed, flight speed, and altitude variations. Specifically, during various maneuvers of an aircraft, it is crucial to assess whether issues such as combustion chamber flameout, intake duct or compressor surge, and overheating of critical components may arise.

The reliability of the engine is paramount as it not only has a direct bearing on flight safety but also influences the maintenance costs associated with the engine. Therefore, achieving high reliability is an essential requirement for ensuring both safety and economic efficiency.

Low operational cost

Cost is a pivotal factor that significantly impacts the economic viability of airlines. The operational cost of aero engines comprises three primary components:

(1) Fuel consumption;

(2) The expenditure associated with engine procurement;

(3) Maintenance expenses.

Manufacturers must take operational cost into account to secure a larger market share.

Often, these requirements for civil aviation engines cannot be simultaneously met, as emphasizing one aspect may inevitably compromise the performance of the others. Consequently, it is imperative to strike a balance among these requirements to optimize overall performance.

 Words & Expressions

1. piston engine	活塞发动机
2. jet engine	喷气式发动机

课文翻译

3. cylinder　　　　　　　　　　　　　　（发动机的）气缸
4. gasoline　　　　　　　　　　　　　汽油
5. aluminum　　　　　　　　　　　　铝
6. cast　　　　　　　　　　　　　　　铸造
7. range　　　　　　　　　　　　　　航程
8. rotary　　　　　　　　　　　　　　旋转的
9. radial　　　　　　　　　　　　　　辐射状的
10. radial engine　　　　　　　　　　　星形发动机
11. supercharging　　　　　　　　　　增压
12. injection　　　　　　　　　　　　　注入，喷射
13. propulsion　　　　　　　　　　　　推进
14. reliability　　　　　　　　　　　　可靠性
15. spin　　　　　　　　　　　　　　　旋转
16. revolutionize　　　　　　　　　　　彻底改变
17. metallurgy [məˈtælədʒi]　　　　　　冶金
18. thrust-to-weight ratio　　　　　　　推重比
19. high-bypass turbofan　　　　　　　高涵道比涡扇发动机
20. civil aviation　　　　　　　　　　　民航
21. mount　　　　　　　　　　　　　　安装
22. turbocharger　　　　　　　　　　　涡轮增压器
23. rotor　　　　　　　　　　　　　　转子
24. stator　　　　　　　　　　　　　　定子，静子叶片
25. booster [ˈbuːstə(r)]　　　　　　　　（电器的）增压机，助推器
26. crystal [ˈkrɪst(ə)l]　　　　　　　　结晶，晶体
27. powder　　　　　　　　　　　　　粉，粉末；粉状物质
28. clearance [ˈklɪərəns]　　　　　　　间距
29. consumption [kənˈsʌmpʃ(ə)n]　　　消耗，消耗量
30. monitor　　　　　　　　　　　　　监视，检查，监控
31. malfunction　　　　　　　　　　　失灵；故障，功能障碍
32. inlet　　　　　　　　　　　　　　进口
33. outlet　　　　　　　　　　　　　　出口
34. frontal　　　　　　　　　　　　　正面的
35. nacelle　　　　　　　　　　　　　机舱，飞机短舱
36. utilization [ˌjuːtəlaɪˈzeɪʃn]　　　　利用，应用，效用

37. payload　　　　　　　　　　　（飞机）有效载荷，装载量
38. compressor surge　　　　　　压气机喘振
39. flameout　　　　　　　　　　熄火

 Exercise

Ⅰ. Fill in the following blanks according to the text.

1. The first 40 years (1903—1945) were the ruling period for _____ engines. The last 80-odd years (1939—now) were the era of jet engines.

2. The resulting hot gases expand and are then exhausted through a turbine, which _____ the compressor and generates thrust.

3. Engines became more reliable, efficient, and durable, thanks to advances in _____, manufacturing processes, and design techniques.

4. Today, aero engines are extremely complex machines that combine advanced technologies with high-performance materials to achieve maximum efficiency and _____.

5. Fuel _____ is the quantity of fuel that enters the combustion chamber per unit time.

Ⅱ. Translate the following sentences into Chinese.

1. The first aero engine in human history was the 12 horsepower 4-cylinder gasoline engine of the Wright brothers in 1903, which used aluminum cast cylinder blocks to solve the problem of reducing weight while maintaining power.

2. Jet engines offer higher speeds, greater efficiency, and improved altitude performance compared to traditional piston engines.

3. Engines are designed to operate at high temperatures and pressures while providing exceptional thrust-to-weight ratios and fuel efficiency.

4. Technologies such as electric propulsion, fuel cells, and alternative fuels hold promise for future aero engines, while specialists will continue to address safety and efficiency concerns as they seek to further push the boundaries of aviation.

5. The engine adopts advanced devices such as wide chord fans, floating wall combustion chambers, single crystal turbine blades and powder metallurgy turbine discs, active control systems of blade tip clearance, and full digital electronic control systems, which enhance the engine's competitiveness.

Module Two　Structure of Aero Engine

As a typical thermal engine, the aero engine is divided into cold end and hot end based on the difference in operating temperature of gases in different parts of the engine. The main components of the cold end include the inlet duct and compressor of the engine, while the main components of the hot end include the combustion chamber, turbine, nozzle, and thrust reverser.

　## Lesson 3　Air Inlet Duct

课文音频

The engine is mounted either within the fuselage of the aircraft or in a dedicated engine nacelle, necessitating an air inlet pipe system to supply the engine with the required airflow. This section of the piping, extending from the inlet of the aero engine to the inlet of the engine compressor, is referred to as the air inlet duct (Figure 3.1). The Mach number of the airflow at the compressor inlet typically remains below 0.7. Therefore, during flight, whenever the flight Mach number surpasses this threshold, the inlet duct performs the crucial function of decelerating the airflow, effectively transforming its kinetic energy into an elevation in pressure. Consequently, the inlet duct is occasionally referred to as an inlet diffuser.

Figure 3.1　Air inlet duct

The primary functions of the air inlet duct are as follows:

(1) To smoothly introduce a sufficient quantity of air into the compressor with minimal flow loss under various conditions.

(2) To enhance the pressure of the air by compressing it through ram pressure when the Mach number of the airflow at the compressor inlet is lower than the flight Mach number.

(3) In all flight conditions and engine operating states, the inlet should avoid excessive time and spatial unevenness of airflow during the pressurization process, in order to reduce the risk of surge in the fan or compressor and vibration of the blades. The external resistance of the inlet should be as small as possible to ensure that the inlet flow field can meet the requirements of the compressor and combustion chamber during normal operation. The flow rate at the inlet of the jet engine must be reduced to $0.3-0.6$ times the local speed of sound, and there are stringent constraints on the uniformity of the inlet flow field.

(4) The inlet of military aircraft must possess "stealth" features, encompassing noise suppression and minimized radar signature. Additionally, when the thrust reverser and thrust vector nozzle are engaged, they must adhere to strict standards for engine exhaust management and ensure maximum protection against the ingress of foreign objects into the engine.

The inlet ducts can be categorized into two major types: subsonic inlet and supersonic inlet. Furthermore, the supersonic inlet ducts can be subdivided into three varieties: internal pressure type, external pressure type, and hybrid type. Currently, China's civil aviation primarily relies on subsonic aircraft, hence the engines typically employ an expanded subsonic inlet design. In essence, the proper functioning of the inlet plays a pivotal role in determining the overall performance of the engine.

The inlet system consists essentially of the inlet duct, inlet duct control device, bleed valve, auxiliary inlet valve, boundary layer suction device, and protective mechanism designed to prevent the inhalation of foreign objects.

3.1 Subsonic Inlet Duct

The subsonic inlet duct comprises a shell and a rectifying cone, and the rectifying cone may be further categorized into a front and a rear rectifying cone. The inlet features a circular lip, and its internal channel is designed as a diverging passage to decelerate and pressurize the airflow, as shown in Figure 3.2.

When the engine is running at a high speed on the ground, the compressor draws air in through the inlet duct, and the pressure of the air at the entrance to the compressor is slightly lower than that of the surrounding or ambient air. When the airplane moves down the runway for takeoff, air is rammed into the duct until its pressure becomes the same as that of the ambient air. The speed at which this happens is called the ram-recovery speed. As the airplane continues to increase its forward speed, the ram effect becomes greater, and even though there is some energy loss due to the increased velocity of the air entering the engine, the thrust increases.

Figure 3.3 shows the type of inlet duct used by a typical high-bypass turbofan. The air flows through the divergent duct into the fan, and although part of the air passes out through the fan discharge, part of it also flows into the low-pressure compressor and supercharges the core engine.

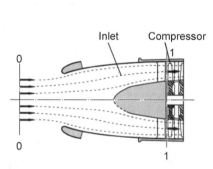
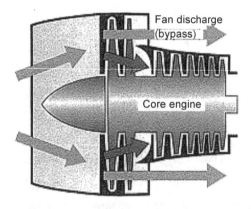

Figure 3.2　Subsonic inlet duct　　　　Figure 3.3　High-bypass turbofan engine

Turbine engines installed in most run-in stands and on some helicopters have bell mouth inlets. These inlets provide a smooth curved surface for the induction air to follow when flowing into the compressor. The duct losses with a bell mouth are extremely low, and this type of inlet is used when calibrating the performance of gas turbine engines.

Blow-in doors

Under some conditions of airflow in a gas turbine engine, the inlet duct doesn't furnish sufficient air to prevent a compressor stall. Installations in which this is a possibility are often equipped with blow-in doors. Blow-in doors are installed in the side of the inlet duct and are spring loaded to hold them closed. But when the inlet air pressure becomes a specified amount lower than that of the ambient air, the pressure differential forces them open and furnishes additional air to the compressor inlet.

Foreign-object damage

Foreign-object damage (FOD) is a major problem with gas turbine engine operation, and when an aircraft is not operating, it is a common practice to install covers over the inlet ducts.

Inlet screens have been tried, but their obstruction of the inlet air, and the probability of their being covered with ice or sucked into the compressor have made them impractical for all but a few of the turboshaft and turboprop engines.

Ice that forms on the lip of the inlet duct can break off and cause damage to the compressor. This problem is minimized by routing warm compressor bleed air between the skins of the inlet duct to prevent the formation of ice.

When air is drawn into the inlet duct, a high-energy vortex is often formed, which creates a strong suction that reaches to the ground from engines mounted low in pods beneath the wing. These vortices can pick up rocks and other debris that, if allowed to enter the engine, will damage the compressor. To prevent this type of damage, vortex dissipaters may be installed below and in front of the inlet duct. These are nozzles that blow a stream of high-velocity compressor bleed air into the area where the vortices form. They force the vortex to dissipate before it can gain enough energy to pick up debris.

3.2 Supersonic inlet Duct

The air approaching the compressor inlet must always be at a speed below the speed of sound. When the aircraft is flying at supersonic speed, the inlet air must be slowed to subsonic speed before it reaches the compressor. This is done by using a convergent-divergent (CD) inlet duct such as the one in Figure 3.4. Air enters the convergent portion of the duct at a supersonic speed, and the velocity decreases until the narrowest part of the duct is reached. At this point the air velocity has been reduced to the speed of sound and a normal shock wave forms. Beyond this point, the duct becomes larger. The air, which has passed through the shock wave, is now flowing at a subsonic speed, and it is further slowed down as it flows through the divergent portion of the duct. By the time it reaches the compressor, its speed is well below the speed of sound, and its pressure has been increased.

High-speed aircraft have variable inlet ducts, and the shape of the airflow changes

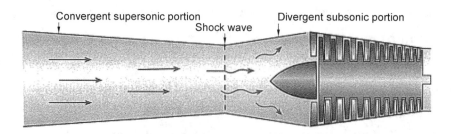

Figure 3.4　A shock wave is formed in the supersonic inlet

with the duct. The change of the duct is done either by lowering or raising a wedge, or by moving a tapered plug in or out of the duct. One or more oblique shock waves form in the duct to slow the air to near the speed of sound, and then a normal shock wave forms to complete the transition from supersonic to subsonic.

Supersonic inlets are divided into three types: internal pressure type, external pressure type, and hybrid type, as shown in Figure 3.5.

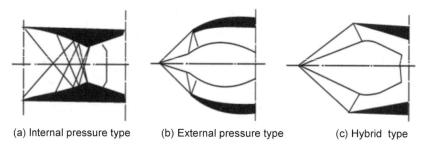

(a) Internal pressure type　　(b) External pressure type　　(c) Hybrid type

Figure 3.5　Supersonic inlet

Internal pressure type: The internal pressure supersonic inlet is designed as a duct that first converges and then diverges. The entire process of decelerating the airflow from supersonic to subsonic speeds takes place within the inlet, leveraging the principle of pressure diffusion to achieve this transition. However, the internal pressure inlet is plagued by a "starting" issue that poses a significant obstacle to its practical deployment.

External pressure type: The external pressure supersonic inlet comprises an outer cover and a central body. As the supersonic airflow passes through the central body, it generates one or multiple oblique shock waves, followed by a final normal shock wave, converting the supersonic airflow into subsonic airflow. Subsequently, the airflow continues to decelerate within the expanding channel. Typically, the external pressure supersonic inlet is restricted to use at flight Mach numbers below 2.0.

Hybrid type: The hybrid supersonic inlet comprises an outer cover and a central body too. Its internal channel, however, resembles the convergent-divergent configuration of the internal pressure supersonic inlet, incorporating both external and internal pressure elements. Once compressed by one or multiple oblique shock waves generated within the outer central body of the inlet, the supersonic airflow maintains its velocity and proceeds into the inlet's interior for further compression. It then traverses the throat or divergent section, ultimately transforming into subsonic flow through a normal shock wave.

Hybrid supersonic inlets, which combine the advantages of both external and internal pressure inlets, are widely employed in aircraft with flight Mach numbers exceeding 2.0.

3.3 Maintenance Procedures and Safety Considerations

The inlet duct is susceptible to damage from external objects, including bird strikes, during operation. Therefore, it is imperative to conduct thorough inspections of areas such as the leading edge lip for signs of damage like pits and cracks. Additionally, routine checks must include a careful examination of fasteners, particularly rivets, on the leading edge of the inlet duct to ensure they are securely tightened. Loose rivets, if detected, should be promptly addressed to prevent them from falling off and potentially causing harm to the engine. Furthermore, the inlet duct's sound attenuation plate may experience corrosion and delamination over time, potentially leading to its detachment and causing severe damage to the engine. Strict adherence to maintenance protocols is crucial to ensure thorough inspection and timely intervention.

When the engine is working, it will inhale a large amount of air, discharge high-temperature and high-speed gas, and generate a lot of noise. To avoid personal injury and equipment damage, dangerous areas are specified in front and behind the aircraft, and their size varies depending on the size, location, thrust, and wind speed of the engine. If you enter the inlet dangerous area, you will be inhaled into the engine. If you enter the exhaust dangerous area, you will be blown away, and the high-temperature exhaust will also cause damage. The size of the dangerous area is related to the power state of the engine, and the larger the power, the larger the range.

Words & Expressions

课文翻译

1. thermal [ˈθɜːm(ə)l]　　　热的
2. component　　　组成部分
3. nozzle　　　喷嘴
4. reverser　　　换向器
5. fuselage　　　(飞机的)机身
6. pipe　　　管道
7. air inlet duct　　　进气道
8. Mach number　　　马赫数
9. vibration　　　振动
10. airflow　　　气流
10. surpass　　　超过
11. kinetic [kɪˈnetɪk]　　　运动的,运动引起的
12. elevation　　　提高,升高
13. sufficient　　　足够的
14. minimal　　　最小的
15. spatial　　　空间的
16. unevenness [ʌnˈiːvnnəs]　　　不均匀
17. in order to　　　为了,目的在于
18. external　　　外部的
19. military　　　军事的
20. stealth [stelθ]　　　隐形的
21. encompass　　　包含,包括
22. suppression　　　抑制
23. ingress [ˈɪŋɡres]　　　进入
24. subsonic　　　亚声速的
25. hybrid [ˈhaɪbrɪd]　　　混合的
26. suction [ˈsʌkʃ(ə)n]　　　吸,吸出
27. inhalation　　　(空气等的)吸入
28. rectifying cone　　　整流锥
29. ambient　　　周围的,周围环境的
30. runway　　　飞机跑道

31. ram 撞击，冲
32. divergent 发散的
33. discharge 排除
34. supercharge 对……增压，涡轮增压
35. helicopter 直升机
36. calibrate ['kælɪbreɪt] 校准
37. stall （飞机的）失速
38. spring 弹簧
39. impede 妨碍，阻碍
40. on the lip of 在……的边缘
41. vortex 涡流，漩涡
42. pod （飞机的）吊舱
43. debris 残骸，碎片
44. dissipate 驱散
45. convergent 收缩的
46. tapered 锥形的
47. oblique （线或角）斜的，倾斜的
48. fastener 紧固件
49. rivet ['rɪvɪt] 铆钉
50. corrosion [kə'rəʊʒ(ə)n] 腐蚀，侵蚀

Exercise

Ⅰ. Fill in the following blanks according to the text.

1. As a typical _____ engine, the aero engine is divided into cold end and hot end based on the difference in operating temperature of gases in different parts of the engine.

2. The engine is mounted either within the _____ of the aircraft or in a dedicated engine nacelle, necessitating an air inlet pipe system to supply the engine with the required airflow.

3. The inlet of _____ aircraft must possess "stealth" features, encompassing noise suppression and minimized radar signature.

4. Currently, China's civil aviation primarily relies on _____ aircraft, hence the engines typically employ an expanded subsonic inlet design.

5. Blow-in doors are installed in the side of the inlet duct and are _____ loaded to hold them closed.

Ⅱ. Translate the following sentences into Chinese.

1. The main components of the cold end include the intake duct and compressor of the engine, while the main components of the hot end include the combustion chamber, turbine, nozzle, and some thrust reversers.

2. Therefore, during flight, whenever the flight Mach number surpasses this threshold, the inlet duct performs the crucial function of decelerating the airflow, effectively transforming its kinetic energy into an elevation in pressure.

3. As the airplane continues to increase its forward speed, the ram effect becomes greater, and even though there is some energy loss due to the increased velocity of the air entering the engine, the thrust increases.

4. But when the inlet air pressure becomes a specified amount lower than that of the ambient air, the pressure differential forces them open and furnishes additional air to the compressor inlet.

5. When air is drawn into the inlet duct, a high-energy vortex is often formed, which creates a strong suction that reaches to the ground from engines mounted low in pods beneath the wing.

6. One or more oblique shock waves form in the duct to slow the air to near the speed of sound, and then a normal shock wave forms to complete the transition from supersonic to subsonic.

 ## Lesson 4　Compressors

课文音频

The compressor is a vital component of aviation gas turbine engines. Its primary function is to compress the airflow passing through it, thereby increasing the pressure of the airflow and creating favorable conditions for gas expansion to perform work, ultimately enhancing the thrust of the engine. The compressor converts mechanical energy from the turbine into kinetic energy in the air. The compressor accelerates the air, which then flows through a diffuser, slowing it down and converting most of the kinetic energy (velocity) into potential energy (pressure) and some into heat. The majority of air flows from the compressor into the combustion section, some of it, called compressor bleed air, is used for anti-icing the inlet ducts and for cooling parts of the hot section. Other bleed air is used for cabin pressurization, air conditioning, fuel system anti-icing, and pneumatic engine starting.

The method used by a compressor to enhance air pressure involves continuously exerting force on the air through a rapidly rotating impeller. Typically, based on the airflow's direction along the compressor rotor and pressurization techniques, as shown in Figure 4.1, compressors can be categorized into centrifugal, axial, and hybrid types. A centrifugal compressor refers to the type where the airflow moves away from the impeller's center, whereas an axial compressor is characterized by airflow parallel to the impeller shaft. A hybrid compressor is a combination of axial and centrifugal compressors. Currently, axial compressors are the most widely utilized.

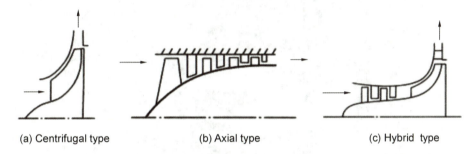

(a) Centrifugal type   (b) Axial type   (c) Hybrid type

Figure 4.1 Classification of compressors

The centrifugal compressor is a single- or two-stage unit employing an impeller to accelerate the air and a diffuser to produce the required pressure rise. The axial compressor is a multi-stage unit employing alternate rows of rotating (rotor) blades and stationary (stator) vanes, to accelerate and diffuse the air until the required pressure rise is obtained.

With regard to the advantages and disadvantages of the two types, the centrifugal compressor is usually more robust than the axial compressor and is also easier to develop and manufacture. The axial compressor however consumes far more air than a centrifugal compressor of the same frontal area and can be designed to attain much higher pressure ratios. Since the air flow is an important factor in determining the amount of thrust, the axial compressor engine will also give greater thrust for the same frontal area. This feature, plus the ability to increase the pressure ratio by addition of extra stages, has led to the adoption of axial compressors in most engine designs. However, the centrifugal compressor is still favoured for smaller engines where its simplicity and ruggedness outweigh any other disadvantages.

4.1　Centrifugal Compressors

The centrifugal compressor consists of an impeller (rotor), a diffuser (stator), and a compressor manifold, as shown in Figure 4.2. Centrifugal compressors have a single or double-sided impeller and occasionally a two-stage, single sided impeller is used. The impeller is supported in a casing that also contains a ring of diffuser vanes. If a double-entry impeller is used, the airflow to the rear side is reversed in direction and a plenum chamber is required.

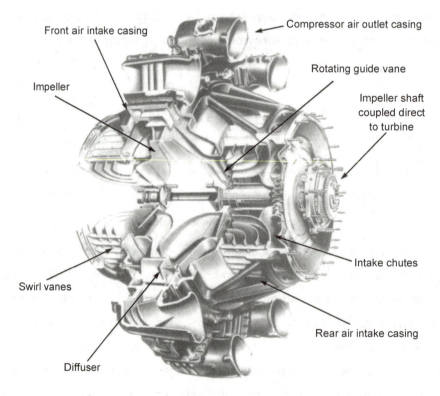

Figure 4.2　A typical centrifugal flow compressor

Centrifugal compressors were used on many of the earliest gas turbine engines because of their ruggedness, light weight, ease of construction, and high pressure ratio for each stage of compression.

Centrifugal compressors have a high pressure rise per stage that can be around 8∶1. Generally centrifugal compressors are limited to two stages due to efficiency concerns. The two main functional elements are the impeller and the diffuser (Figure 4.3). Although the diffuser is a separate unit and is placed inside and bolted to the manifold, the entire assembly (diffuser and manifold) is often referred to as the diffuser. The impeller is usually made of forged aluminum alloy, and is heat treated, machined, and smoothed for minimum flow restriction and turbulence.

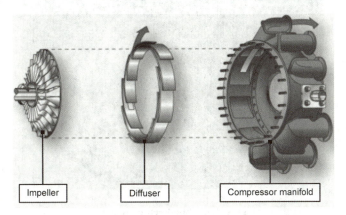

Figure 4.3　Components of a centrifugal compressor

Impellers

The impeller consists of a forged disc with integral, radially disposed vanes on one or both sides (Figure 4.4) forming convergent passages in conjunction with the compressor casing. The vanes may be swept back, but for ease of manufacture, straight radial vanes are usually employed. In order to facilitate the axial flow of air from the entry duct on to the rotating impeller, the vanes in the centre of the impeller are curved along the direction of rotation. The curved sections may be integral with the radial vanes or formed separately for easier and more accurate manufacture. The impeller draws in air and accelerates it outward by centrifugal force. Centrifugal compressors can have one or two impellers. Compressors having only one impeller are referred to as single-stage compressors, those with two impellers are referred to as double-stage compressors. More than two compressor stages is impractical because the benefits gained by additional stages are negated by energy losses when the airflow slows between impellers. Furthermore, due to the added weight of each impeller, more energy are required from an engine to drive the compressor.

Diffusers

The diffuser is an annular chamber provided with a number of vanes forming a series of divergent passages into the manifold. The diffuser assembly may be an integral part of the compressor casing or a separately attached assembly. The diffuser vanes direct the flow of air from the impeller to the manifold at an angle designed to obtain the maximum pressure energy of the airflow. The diffuser blade consists of a number of vanes formed tangential to the impeller. The vane passages are divergent to

Figure 4.4　Single-entry impeller

convert the kinetic energy into pressure energy and the inner edges of the vanes are in line with the direction of the resultant airflow from the impeller (Figure 4.5). The clearance between the impeller and the diffuser is an important factor, as too small a clearance will set up aerodynamic buffeting impulses that could be transferred to the impeller and create an unsteady airflow and vibration.

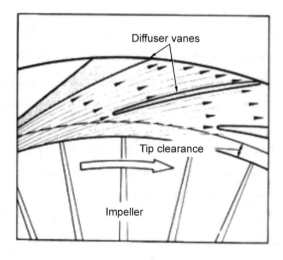

Figure 4.5　Airflow at entry to diffuser

Compressor manifold

The compressor manifold shown in Figure 4.3 diverts the flow of air from the diffuser, which is an integral part of the manifold, into the combustion chambers. The manifold has one outlet port for each chamber so that the air is evenly divided. A compressor outlet elbow is bolted to each of the outlet ports. These air elbows are constructed in the form of ducts and are known by a variety of names, such as air outlet ducts, air outlets, or combustion chamber inlet ducts. Regardless of the terminology used, these outlet elbows perform a very important part of the diffusion process, that is, they change the direction of the airflow from radial to axial, and the gas diffusion process is completed after the turn. To help the elbows perform this function in an efficient manner, turning vanes (cascade vanes) are sometimes fitted inside the elbows. These vanes reduce air pressure losses by providing a smooth, turning surface.

When the impeller rotates at a high speed, air is continuously induced into the centre of the impeller. Centrifugal action causes the air to flow radially outwards along the vanes to the impeller tip, thus accelerating the air and also causing a rise in pressure. The engine inlet duct may contain vanes that provide an initial swirl to the air entering the compressor. The air, on leaving the impeller, passes into the diffuser section where the passages form divergent nozzles that convert most of the kinetic energy into pressure energy, as illustrated in Figure 4.6. In practice, it is usual to design the compressor so that about half of the pressure rise occurs in the impeller and half in the diffuser. To maximize the airflow and pressure rise through the compressor, high speed rotation of the impeller is required, therefore the impeller is designed to operate at tip speeds of up to 1,600 feet per sec. By operating at such high tip speeds the air velocity from the impeller is increased so that greater energy is available for conversion into pressure. To maintain the efficiency of the compressor, it is necessary to keep the clearance between the impeller and the casing as small as possible to prevent excessive air leakage (Figure 4.7).

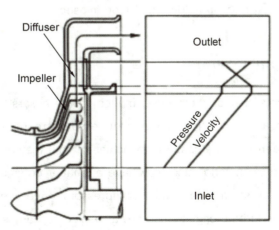

Figure 4.6 Pressure and velocity changes through a centrifugal compressor

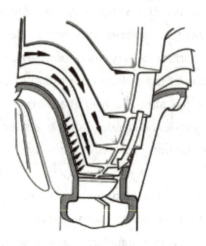

Figure 4.7 Impeller working clearance and air leakage

Included in the ducting of double-entry compressor engines is the plenum chamber. This chamber is necessary for a double-entry compressor because the air must enter the engine at almost a right angle to the engine axis. Therefore, in order to give a positive flow, the air must surround the engine compressor at a positive pressure before entering the compressor. Included in some installations as necessary parts of the plenum chamber are the auxiliary blow-in doors (air-intake doors). These blow-in doors bring air into the engine compartment during ground operation, when the air demand of the engine is in excess of the airflow through the inlet ducts. The doors are held closed by spring action when the engine is not operating. During operation, the doors open automatically whenever engine compartment pressure drops below atmospheric pressure. During takeoff and flight, ram air pressure in the engine compartment aids the springs in holding the doors closed.

4.2 Axial Compressors

Axial compressors are, as their name implies, compressors through which the air passes axially or in a straight line. Axial compressors are heavier than centrifugal compressors and are much more costly to manufacture, but they are capable of a much higher overall compression ratio, and they have a smaller cross-sectional area which makes them easier to streamline. Axial compressors have therefore become the standard for large gas turbine engines and are also used in many small engines.

The axial compressor has two main elements: a rotor and a stator. The rotor has blades fixed on a spindle. These blades impel air rearward in the same manner as a propeller because of their specific angle and airfoil contour. The rotor, turning at high speed, takes in air at the compressor inlet and impels it through a series of stages. From inlet to exit, the air flows along an axial path and is compressed at a ratio of approximately 1.25 : 1 in each stage. The action of the rotor increases the air compression rate at each stage and accelerates the airflow rearward through several stages. With this increased velocity, energy is transferred from the compressor to the air. The stator blades act as diffusers at each stage, partially converting high velocity into pressure. Each consecutive pair of rotor and stator blades constitutes a pressure stage. The number of rows of blades (stages) is determined by the amount of air and total pressure rise required. Compressor pressure ratio increases with the number of compression stages. Most engines utilize up to 16 stages or more.

Rotors

In compressor designs (Figure 4.8) the rotational speed is such that a disc is required to support the centrifugal blade load. Where a number of discs are fitted onto one shaft they may be coupled and secured together by a mechanical fixing but generally the discs are assembled and welded together, close to their periphery, thus forming an integral drum. Typical methods of securing rotor blades to the disc are shown in Figure 4.9. The rotor blades can be fixed circumferential or axial to suit special requirements of the stage. In general the aim is to design a securing feature that imparts the lightest possible load on the supporting disc thus minimizing the disc's weight. Whilst most compressor designs have separate blades for manufacturing and maintainability requirements, it becomes more difficult on the smallest engines to design a practical fixing. However this may be overcome by producing discs integrated with blades, i.e. the so called blisk.

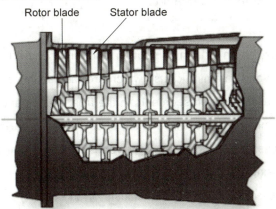

Figure 4.8 Disk-type compressor rotor

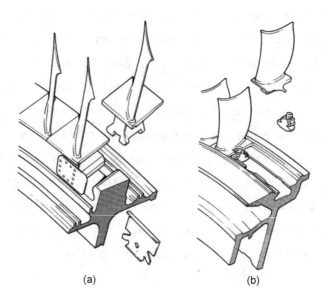

Figure 4.9 Methods of securing blades to disc

Rotor blades

Rotor blades are of airfoil section and usually designed to give a pressure gradient along their length to ensure that the air maintains a reasonably uniform axial velocity. The higher pressure towards the tip balances out the centrifugal action of the rotor on the airstream. To obtain these functions, it is necessary to "twist" the blade from root to tip to give the correct angle of incidence at each point. Air flowing through a compressor creates two boundary layers on the inner and outer walls, in which the air flows slowly until it comes to a standstill. In order to compensate for the slow air in the boundary layer, a localized increase in blade camber both at the blade tip and root has been introduced. The blade extremities appear as if they are formed by bending over each corner, hence the term "end-bend".

Stator vanes

Stator vanes are again of airfoil section and are secured into the compressor casing or into stator vane retaining rings, which are themselves secured to the casing. The vanes are often assembled in segments in the front stages and may be shrouded at their inner ends to minimize the vibrational effect of flow variations on the longer vanes. It is also necessary to lock the stator vanes in such a manner that they will not rotate around the casing.

An axial compressor (Figure 4.10) consists of one or more rotor assemblies that carry blades of airfoil section. These assemblies are mounted between bearings in the casings which incorporate the stator vanes. The compressor is a multi-stage unit as the amount of pressure increase provided by each stage is small; a stage consists of a row of rotating blades followed by a row of stator vanes. When several stages of compression are connected in series on one shaft, it becomes necessary to vary the stator vane angle to enable the compressor to operate effectively at speeds below the design speed. As the pressure ratio is increased the incorporation of variable stator vanes ensures that the airflow is directed onto the succeeding stage of rotor blades at an acceptable angle.

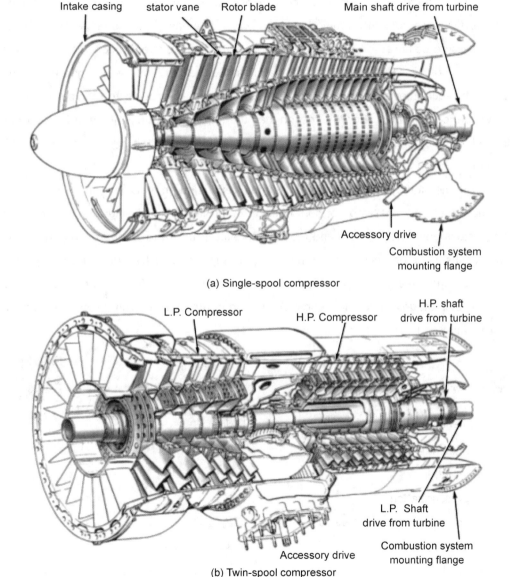

Figure 4.10 Typical axial compressors

A single-spool compressor (Figure 4.10a) consists of one rotor assembly and stators with as many stages as necessary to achieve the desired pressure ratio and all the airflow from the inlet passes through the compressor. The multi-spool compressor consists of two or more rotor assemblies, each driven by their own turbine at an optimum speed to achieve higher pressure ratios and to give greater operating flexibility. Although a twin-spool compressor (Figure 4.10b) can be used for a pure jet engine, it is most suitable for the bypass type of engine where the front or low pressure compressor is designed to handle a larger airflow than the high pressure compressor. Only a percentage of the air from the low pressure compressor passes into the high pressure compressor, the remainder of the air, the bypass flow, is ducted around the high pressure compressor. Both flows mix in the exhaust system before passing to the propelling nozzle. This arrangement matches the velocity of the jet nearer to the optimum requirements of the aircraft and results in higher propulsive efficiency, hence lower fuel consumption. For this reason the pure jet engine where all the airflow passes through the full compression cycle is now obsolete for all but the highest speed aircraft engine.

With the high bypass turbofan engine this trend is taken a stage further. The intake air undergoes only one stage of compression in the fan before being split between the core or gas generator system and the bypass duct in the ratio of approximately 1∶5. This results in the optimum arrangement for passenger and/or transport aircraft flying at just below the speed of sound. The fan may be coupled to the front of a number of core compression stages (two-shaft engine) or a separate shaft driven by its own turbine (three-shaft engine).

 Words & Expressions

课文翻译

1. mechanical　　　　　　　　　机械的
2. pneumatic　　　　　　　　　气动的
3. impeller　　　　　　　　　　叶轮
4. pressurization　　　　　　　　增压
5. centrifugal [ˌsentrɪˈfjuːg(ə)l]　　离心的
6. casing [ˈkeɪsɪŋ]　　　　　　　框，壳
7. axial　　　　　　　　　　　　轴向的
8. robust　　　　　　　　　　　结实的，坚固的

9. manufacture　　　　　　　　　制造
10. consume　　　　　　　　　　消耗
11. frontal area　　　　　　　　最大截面
12. favoured　　　　　　　　　　受优惠的,有特权的
13. manifold　　　　　　　　　　集气管,多支管
14. vane　　　　　　　　　　　　叶片
15. plenum chamber　　　　　　增压室
16. forge　　　　　　　　　　　锻造
17. alloy　　　　　　　　　　　合金
18. turbulence ['tɜːbjələns]　　　　湍流
19. centrifugal force　　　　　　离心力
20. negate　　　　　　　　　　　使无效,否定
21. annular　　　　　　　　　　环状的
22. right angle　　　　　　　　直角
23. cross-sectional area　　　　横截面面积
24. streamline　　　　　　　　　把……做成流线型
25. spindle　　　　　　　　　　轴
26. impel　　　　　　　　　　　推动
27. consecutive [kən'sekjətɪv]　连续的,不间断的
28. weld　　　　　　　　　　　焊接
29. periphery [pə'rɪfəri]　　　边缘,外围
30. circumferential [səˌkʌmfə'renʃəl]　圆周的
31. disc　　　　　　　　　　　圆盘状物
32. blisk　　　　　　　　　　整体叶盘
33. airfoil ['eəfɔɪl]　　　　翼面
34. gradient　　　　　　　　梯度
35. airstream　　　　　　　气流
36. stagnant　　　　　　　　不流动的,停滞的
37. segment　　　　　　　　分段
38. flexibility　　　　　　灵活性
39. remainder　　　　　　　剩余物
40. optimum　　　　　　　　最适宜的,最佳的

Exercise

Ⅰ. Fill in the following blanks according to the text.

1. The compressor converts mechanical energy from the turbine into _____ energy in the air.

2. The method used by a compressor to enhance air pressure involves continuously exerting force on the air through a rapidly rotating _____.

3. The centrifugal compressor consists of an impeller (rotor), a diffuser (stator), and a compressor _____.

4. The impeller is usually made of forged aluminum _____, and is heat treated, machined, and smoothed for minimum flow restriction and turbulence.

5. The diffuser is an _____ chamber provided with a number of vanes forming a series of divergent passages into the manifold.

6. Rotor blades are of _____ section and usually designed to give a pressure gradient along their length to ensure that the air maintains a reasonably uniform axial velocity.

Ⅱ. Translate the following sentences into Chinese.

1. The majority of air flows from the compressor into the combustion section, some of it, called compressor bleed air, is used for anti-icing the inlet ducts and for cooling parts of the hot section. Other bleed air is used for cabin pressurization, air conditioning, fuel system anti-icing, and pneumatic engine starting.

2. A centrifugal compressor refers to the type where the airflow moves away from the impeller's center, whereas an axial compressor is characterized by airflow parallel to the impeller shaft.

3. Centrifugal compressors were used on many of the earliest gas turbine engines because of their ruggedness, light weight, ease of construction, and high pressure ratio for each stage of compression.

4. Regardless of the terminology used, these outlet elbows perform a very important part of the diffusion process; that is, they change the direction of the airflow from radial to axial, and the gas diffusion process is completed after the turn.

5. The vanes are often assembled in segments in the front stages and may be shrouded at their inner ends to minimize the vibrational effect of flow variations on the longer vanes.

Lesson 5　Combustion Section

课文音频

　　The combustion section houses the combustion process, which raises the temperature of the air passing through the engine. This process releases energy contained in the air-fuel mixture. The major part of this energy is required at the turbine or turbine stages to drive the compressor. About 2/3 of the energy is used to drive the gas generator compressor. The remaining energy passes through the remaining turbine stages that absorb more of the energy to drive the fan, output shaft, or propeller. Only the pure turbojet allows the air to create all the thrust or propulsion by exiting the rear of the engine in the form of a high-velocity jet. These other engine types have some high-speed jet discharged at the rear of the engine but most of the thrust or power is generated by the additional turbine stages driving a large fan, propeller, or helicopter rotor blades.

　　The combustion chamber (Figure 5.1) has the difficult task of burning large quantities of fuel, supplied through the fuel spray nozzles, with extensive volumes of air, supplied by the compressor, and releasing heat in such a manner that the air is expanded and accelerated to give a smooth stream of uniformly heated gas at all conditions required by the turbine. This task must be accomplished with the minimum loss in pressure and with the maximum heat release for the limited space available.

　　The primary function of the combustion section is, of course, to burn the air-fuel mixture, thereby adding heat energy to the air. In order to do this efficiently, there are several very stringent requirements for the combustors used in a turbine engine. Some of these are:

　　(1) minimum pressure loss in gases as they pass through the combustor;

　　(2) high combustion efficiency, therefore low smoke emission;

　　(3) low risk of flame blowout;

　　(4) combustion occurring entirely within the combustor;

　　(5) uniform temperature distribution throughout the gases;

　　(6) low enough temperature of the gases leaving the combustor to prevent damage to the turbine;

　　(7) combustor design providing easy starting.

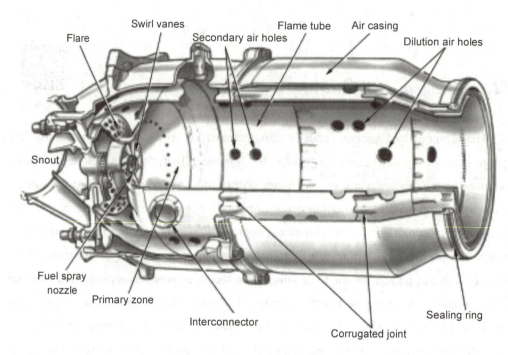

Figure 5.1 An early combustion chamber

The location of the combustion section is directly between the compressor and the turbine sections. The combustion chambers are always arranged coaxially with the compressor and turbine regardless of type, since the chambers must be in a through-flow position to function efficiently. All combustion chambers contain the same basic elements: casing, perforated inner liner, fuel injection system, some means for initial ignition and fuel drainage system to drain off unburned fuel after engine shutdown.

5.1 Combustion Process

Air from the engine compressor enters the combustion chamber at a velocity up to 500 feet per second, but this air speed is far too high for combustion, so the first thing that the chamber must do is to diffuse it, i.e. decelerate the air and raise its static pressure. Since the speed of burning kerosine at normal mixture ratios is only a few feet per second, any fuel lit even in the diffused air stream, which now has a velocity of about 80 feet per second, would be blown away. A region of low axial velocity has therefore to be created in the chamber, so that the flame will remain alight throughout the range of engine operating conditions.

In normal operation, the ratio of air to fuel in the overall combustion chamber can vary between 45 : 1 and 130 : 1. However, kerosine will only burn efficiently at, or close to, a ratio of 15 : 1, so the fuel must be burned with only part of the air entering the chamber, in what is called a primary combustion zone. Approximately 20 per cent of the air mass flow is taken in by the snout or entry section (Figure 5.2).

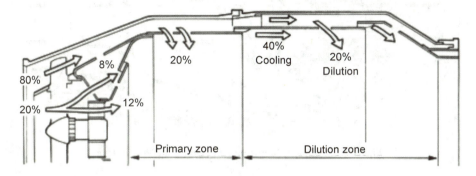

Figure 5.2 Apportioning the airflow

Immediately downstream of the snout are swirl vanes and a perforated flare, through which air passes into the primary combustion zone. The swirling air induces a flow upstream of the centre of the flame tube and promotes the desired recirculation. The air not picked up by the snout flows into the annular space between the flame tube and the air casing. The air from the swirl vanes and that from the secondary air holes interacts and creates a region of low velocity recirculation. This low velocity recirculation takes the form of a toroidal vortex, similar to a smoke ring, which has the effect of stabilizing and anchoring the flame (Figure 5.3). The recirculating gases hasten the burning of freshly injected fuel droplets by rapidly bringing them to ignition temperature. It is arranged that the conical fuel spray from the nozzle intersects the recirculation vortex at its centre. This action, together with the general turbulence in the primary zone, greatly assists in breaking up the fuel and mixing it with the incoming air.

The temperature of the gases released by combustion is about 1,800 to 2,000 deg. C., which is far too hot for entry to the nozzle guide vanes of the turbine. The air not used for combustion, which amounts to about 60 per cent of the total airflow, is therefore introduced progressively into the flame tube. Approximately a third of this is used to lower the gas temperature in the dilution zone before it enters the turbine and the remainder is used for cooling the walls of the flame tube. This is achieved by a film of cooling air flowing along the inside surface of the flame tube wall, insulating it from the hot combustion gases.

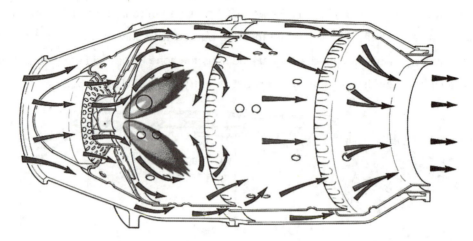

Figure 5.3 Flame stabilizing and general airflow pattern

5.2 Types of Combustion Chamber

There are three main types of combustion chamber in use for gas turbine engines. These are the multiple-can combustion chamber, the can-annular combustion chamber and the annular combustion chamber. Combustors may also be of the straight-through or reverse-flow type.

Multiple-can combustion chamber

The multiple-can combustion chamber is used in centrifugal compressor engines and the earlier types of axial compressor engines. It is a direct development of the early type of Whittle combustion chamber. The major difference is that the Whittle chamber has a reverse flow, which causes a considerable pressure loss.

Air enters the left end of the combustor and divides into two streams. The primary air, which makes up about one fourth of the mass and enters into the combustion process, flows into the inner liner where its velocity is slowed down enough that it will not blow out the flame of the burning fuel. The remainder of the air, the secondary air, flows between the liners, part of it enters the inner liner downstream of the combustion in a turbulent fashion so it will mix with the primary air to lower its temperature enough that it will not damage the turbine. There are usually 8 − 10 can-type combustors arranged around the outside of the turbine shaft between the compressor manifold and the turbine.

Igniters are normally installed in only two of the cans to provide the spark for starting the engine. The cans are connected by small flame propagation tubes, also called crossover tubes, so that the flames from the cans containing the igniters are carried to the other cans for ignition. The advantage of can-type combustors is that individual cans may be removed for inspection and replacement without disturbing the others. But they have the disadvantage that the turbine is subjected to uneven temperatures, and if one can fails, the resulting extreme temperature difference can cause turbine failure.

Can-annular combustion chamber

Can-annular combustion chambers are used in many large turbojet and turbofan engines. They consist of individual cans into which fuel is sprayed and ignited. They mount on an annular duct through which hot gases from individual cans are collected and directed uniformly into the turbine.

Figure 5.4 shows a typical can-annular combustion chamber. Fuel is sprayed into the combustor from the fuel manifold and then discharged through the hole in the end of the can. It mixes with air from the compressor and is burned. Secondary air flows through the holes in the can and cools the air before it passes through the turbine. The cutaway in Figure 5.4 shows the perforated tube that carries additional air for cooling. The cans are shorter than those used in a multiple-can combustor and therefore the gas has a lower pressure drop when passing through them. Combining the gases from all of the cans provides a uniform temperature at the turbine, even if one of the fuel nozzles becomes clogged.

Annular combustion chamber

Many modern axial-compressor engines use a single annular combustion chamber. Annular combustion chambers make the most efficient use of the limited space available , and they provide exceptionally efficient mixing of the fuel vapor with the air. They require a minimum amount of cooling air to keep the temperature low enough to prevent damage to the turbine, and they provide an even temperature distribution of the air leaving the combustor. The main disadvantage of an annular combustion chamber is that it cannot be replaced without removing the engine from the aircraft.

In engines where the overall length is critical, some manufacturers have chosen to use reverse-flow combustion chambers such as the one used in the Pratt & Whitney of Canada FT6. The air enters the engine from the rear and flows forward through the axial compressor, then through the centrifugal compressor and diffuser. It flows around the

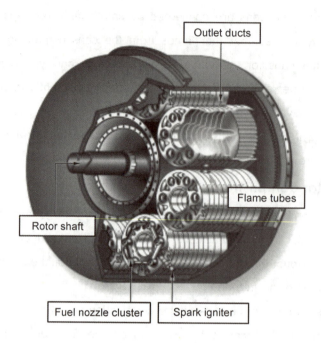

Figure 5.4 A typical can-annular combustion chamber

outside of the combustor and reverses direction to enter the combustor and flows rearward. Fuel is sprayed in and burned at the forward end of the combustor. Secondary air flows in through the slots and holes in the inner liner of the combustor to dilute and cool the air before it leaves. The air then makes its final reverse in direction and flows forward through the compressor turbine and the propeller turbine, which remove most of its energy. The hot air, with the greater part of its energy removed, leaves the engine near the front through the exhaust stacks.

5.3 Maintenance and Safety Notes for Typical Engine Combustors

Common failures encountered in combustion chambers are localized overheating and flameout. In the primary combustion zone, the flame barrel walls and internal components of the chamber must be capable of resisting extremely high gas temperatures. This is typically achieved in practice through the use of superior heat-resistant materials, specifically by applying a high heat-resistant coating and flame retardant insulation to cool the inner walls of the flame cartridges. Furthermore, the combustion chamber must also withstand corrosion caused by combustion products, as well as creep failures that may arise due to temperature variations and fatigue vibration stresses. Localized overheating can result in excessive temperature variations throughout the flame tube, ultimately leading to deformation or the formation of cracks. This phenomenon is primarily attributed to factors such as uneven fuel distribution and disrupted air flow.

(1) The potential reason for uneven fuel distribution lies in changes in the nozzle aperture, such as carbon deposition and rusting, which can lead to excessively large injection angles. Additionally, poor fuel atomization quality or a slanted nozzle installation can result in a deflected fuel injection cone.

(2) Potential reasons for disrupted air flow include flame tube deformation, incorrect installation of the flame tube, and excessive fuel injection.

To prevent damage to the flame tube caused by localized overheating in the combustion chamber, it is crucial to enhance the maintenance of the injector nozzle, ensure proper disassembly and assembly of the combustion chamber, and operate the engine as per regulations, avoiding any sudden surge in fuel injection. Flameout can be categorized into lean flameout and rich flameout, both ultimately stemming from the excess gas coefficient exceeding the range of stable combustion. To mitigate the risk of flameout, it is imperative to utilize aircraft and engines according to specified procedures and perform operations gently. Furthermore, regular inspections of the anti-surge system should be conducted to maintain its optimal condition, thereby preventing flameout and engine shutdown due to surge.

(1) The combustion chamber, being the zone where the oil-gas mixture ignites, constantly withstands high-temperature ablation. During routine borescope inspections, if severe local ablation is observed, it is imperative to assess the uniformity of fuel nozzle atomization in that specific area, as this could lead to localized overheating of the combustion chamber liner.

(2) It is commonplace for excess fuel to accumulate within the combustion chamber during wet and cold engine operations, potentially leaking out through the chamber casing installation edges. Prior to normal engine startup, it is crucial to adhere to the manual's instructions and perform a dry cold run to purge the accumulated fuel, thereby preventing engine damage.

(3) Bore inspection of the combustion chamber is strictly confined to visually accessible areas. This inspection can be conducted through boreholes and ignition

nozzle mounts, utilizing either a rigid or flexible endoscope. The inspection areas within the combustion chamber include the flame tube nose region, as well as the inner and outer flame tube zones.

Words & Expressions

课文翻译

1. combustion	燃烧
2. flame blowout	熄火
3. distribution	分布
4. coaxially	同轴地
5. perforate	穿孔于,在……上打眼
6. injection	注入,喷射
7. drainage	排水,放水;排水系统
8. kerosine	航空燃油
9. snout	锥形进口
10. swirl vanes	旋流器
11. toroidal	环形的
12. dilution	稀释物
13. igniter	点火器,电嘴
14. propagation	传播,传输
15. spray	喷
16. clog	阻塞,堵塞
17. exhaust	消耗,耗尽
18. stack	大量
19. flame cartridge	火焰筒
20. atomization	雾化
21. deformation	变形
22. mitigate	减轻
23. purge	清除,(使)净化
24. bore inspection	孔探检查
25. endoscope	内窥镜

Exercise

Ⅰ. **Fill in the following blanks according to the text.**

1. The combustion section houses the _____ process, which raises the temperature of the air passing through the engine.

2. The combustion chambers are always arranged _____ with the compressor and turbine regardless of type, since the chambers must be in a through-flow position to function efficiently.

3. Immediately downstream of the snout are swirl vanes and a perforated _____, through which air passes into the primary combustion zone.

4. The cans are connected by small flame _____ tubes, also called crossover tubes, so that the flames from the cans containing the igniters are carried to the other cans for ignition.

5. Additionally, poor fuel _____ quality or a slanted nozzle installation can result in a deflected fuel injection cone.

Ⅱ. **Translate the following sentences into Chinese.**

1. These other engine types have some high-speed jet charged at the rear of the engine but most of the thrust or power is generated by the additional turbine stages driving a large fan, propeller, or helicopter rotor blades.

2. Since the speed of burning kerosine at normal mixture ratios is only a few feet per second, any fuel lit even in the diffused air stream, which now has a velocity of about 80 feet per second, would be blown away.

3. This low velocity recirculation takes the form of a toroidal vortex, similar to a smoke ring, which has the effect of stabilizing and anchoring the flame.

4. The primary air, which makes up about one fourth of the mass and enters into the combustion process, flows into the inner liner where its velocity is slowed down enough that it will not blow out the flame of the burning fuel.

5. But they have the disadvantage that the turbine is subjected to uneven temperatures, and if one can fails, the resulting extreme temperature difference can cause turbine failure.

Lesson 6　Turbine Section

The turbine has the task of providing the power to drive the compressor and accessories and, in the case of engines which do not make use solely of a jet for propulsion, of providing shaft power for a propeller or rotor. It completes these tasks by extracting energy from the hot gases released from the combustion system and expanding them to a lower pressure and temperature. High stresses are involved in this process, and for efficient operation, the turbine blade tips may rotate at speeds over 1,500 feet per second. The continuous flow of gas to which the turbine is exposed may have an entry temperature between 850 °C and 1,700 °C and may reach a velocity of over 2,500 feet per second in parts of the turbine.

After the fuel-air mixture burns in the combustor, its energy must be extracted. A turbine transforms a portion of the kinetic energy in the hot exhaust gases into mechanical energy to drive the compressor and accessories. In a turbojet engine, the turbine absorbs approximately 60% to 80% of the total pressure energy from the exhaust gases. The exact amount of energy absorption by the turbine is determined by the load the turbine is driving (i.e., the size and type of the compressor, the number of accessories, and the load applied by the other turbine stages).

To produce the driving torque, the turbine may consist of several stages, each employing one row of stationary nozzle guide vanes and one row of moving blades (Figure 6.1). The number of stages depends upon the relationship between the power required from the gas flow, the rotational speed at which it must be produced and the diameter of turbine permitted.

The turbine assembly consists of two basic elements: turbine inlet guide vane and turbine disk. The turbine inlet guide vane is a stator element, which is also known as the turbine inlet nozzle vane and nozzle diaphragm. The turbine inlet nozzle vanes are located directly aft of combustion chambers and immediately forward of the turbine wheel. This is where the temperature is highest in contact with metal components in the engine. The turbine inlet temperature must be controlled, or damage will occur to the turbine inlet vanes.

The mean blade speed of a turbine has a considerable effect on the maximum

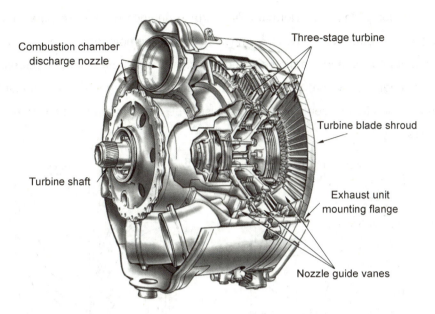

Figure 6.1　T triple-stage turbine with single shaft system

efficiency possible for a given output. For a given output, the gas velocities, deflections, and hence losses, are reduced in proportion to the square of mean blade speeds. Stress in the turbine disc increases as the square of the speed, therefore to maintain the same stress level at higher speed, the sectional thickness, hence the weight, must be increased disproportionately. For this reason, the final design is a compromise between efficiency and weight. Engines operating at higher turbine inlet temperatures are thermally more efficient and have an improved power to weight ratio. Bypass engines have a better propulsive efficiency and thus can have a smaller turbine for a given thrust.

　　The design of the nozzle guide vane and turbine blade passages is based broadly on aerodynamic considerations, and to obtain optimum efficiency, compatible with compressor and combustion design, the nozzle guide vanes and turbine blades are of a basic aerofoil shape. There are three types of turbine: impulse type, reaction type and a combination of the two known as impulse-reaction type. In the impulse type, the total pressure drop across each stage occurs in the fixed nozzle guide vanes which, because of their convergent shape, increase the gas velocity whilst reducing the pressure. The gas is directed onto the turbine blades which experience an impulse force caused by the impact of the gas on the blades. In the reaction type, the fixed nozzle guide vanes are designed to alter the direction of the gas flow without changing the pressure. The converging blade passages experience a reaction force resulting from the expansion and acceleration of the gas. Normally gas turbine engines do not use pure impulse or pure reaction turbine blades but the impulse-reaction combination (Figure 6.2). The proportion of impulse and reaction set in the design of a turbine is largely dependent on the type of engine in which the turbine is to operate, but in general it is about 50 per cent impulse and 50 per cent reaction. Impulse-type turbines are used for cartridge and air starters.

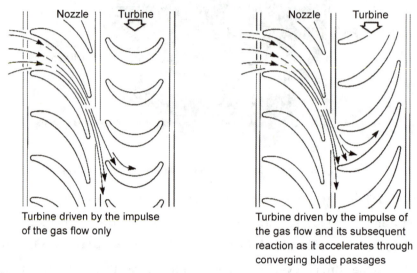

Figure 6.2 Comparison between a pure impulse turbine and an impulse-reaction turbine

6.1 Construction of turbine

The basic components of the turbine are the combustion discharge nozzles, the nozzle guide vanes, the turbine discs and the turbine blades. The rotating assembly is carried on bearings mounted in the turbine casing and the turbine shaft may be common to the compressor shaft or connected to it by a self-aligning coupling.

Nozzle guide vanes

The nozzle guide vanes are of an aerofoil shape with the passage between adjacent vanes forming a convergent duct. The vanes are located in the turbine casing (Figure 6.3) in a manner that allows for expansion.

The nozzle guide vanes are usually of hollow form and may be cooled by passing compressor delivery air through them to reduce the effects of high thermal stresses and gas loads. Turbine discs are usually manufactured from a machined forging with an integral shaft or with a flange onto which the shaft may be bolted. The disc also has, around its perimeter, provision for the attachment of the turbine blades. To limit the effect of heat conduction from the turbine blades to the disc, a flow of cooling air is passed across both sides of each disc.

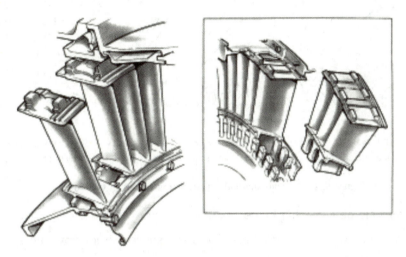

Figure 6.3　Typical nozzle guide vanes showing their shape and location

Turbine blades

The turbine blades are of an aerofoil shape, designed to provide passages between adjacent blades that give a steady acceleration of the flow up to the "throat", where the area is smallest and the velocity reaches that required at the exit to produce the required reaction. The actual area of each blade cross-section is fixed by the permitted stress in the material used and by the size of any holes which may be required for cooling purposes. High efficiency demands thin trailing edges to the sections, but a compromise has to be made so as to prevent the blades cracking due to the temperature changes during engine operation.

The method of attaching the blades to the turbine disc is of considerable importance, since the stress in the disc around the fixing or in the blade root has an important bearing on the limiting rim speed. The blades on the early Whittle engine were attached by the de Laval bulb root fixing, but this design was soon superseded by the "fir-tree" fixing that is now used in the majority of gas turbine engines. This type of fixing involves very accurate machining to ensure that the loading is shared by all the serrations. The blade is free in the serrations when the turbine is stationary and is stiffened in the root by centrifugal loading when the turbine is rotating. Various methods of blade attachment are shown in Figure 6.4; however, the BMW hollow blade and the de Laval bulb root types are now not generally used on gas turbine engines.

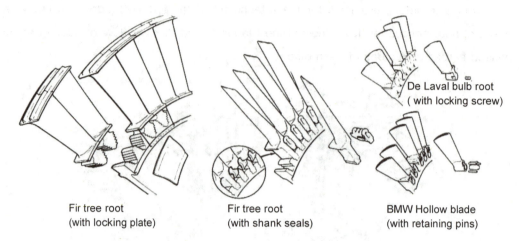

Figure 6.4　Various methods of attaching blades to turbine discs

A gap exists between the blade tips and casing, which varies in size due to the different rates of expansion and contraction. To reduce the loss of efficiency through gas leakage across the blade tips, a shroud is often fitted as shown in Figure 6.1. The shroud is made up by a small segment at the tip of each blade which forms a peripheral ring around the blade tips. An abradable lining in the casing may also be used to reduce gas leakage. Active clearance control (ACC) is a more effective method of maintaining minimum tip clearance throughout the flight cycle. Air from the compressor is used to cool the turbine casing and when used with shroudless turbine blades, enables higher temperatures and speeds to be used.

6.2　Turbine Failures

The turbine section of a gas turbine engine operates in a most hostile environment and is the portion of the engine that requires the most careful inspection. Three main causes of turbine failure are creep, metal fatigue, and corrosion.

Creep is the deformation of a metal part that is continually exposed to high centrifugal loads and high temperature. When the metal is operated within its elastic limit, it will always return to its original configuration when the stress is removed; but if the load is increased, part of the deformation will remain after the stress is removed. Some metals show a permanent deformation even after exposure to a lower amount of stress, if the load is maintained over a long period of time. This type of deformation is called creep.

Metal fatigue is a weakening of the metal that is subjected to repeated cyclic loading. Each time a gas turbine engine is started, operated through its flight routine, and shut down, it goes through a severe heat cycle. The numerous heat cycles that an engine goes through make the metal parts subject to fatigue.

Corrosion is the electrolytic action that occurs when the alloying agents in the metal combine with elements in the air to form salts that have no strength. The corrosive action is accelerated when the metal parts are exposed to extremely high temperatures.

6.3 Turbine Cooling

The primary factor that limits the amount of power a gas turbine engine can produce is the maximum temperature that can be tolerated at the turbine inlet. This is called the turbine inlet temperature (TIT). Some modern engines have increased fuel efficiency because of increased allowable TIT, and one way to increase the allowable TIT is to cool the turbine inlet guide vanes and the first-stage rotor blades.

Turbine cooling is done by flowing compressor bleed air through hollow guide vanes and rotor blades. Air leaves the blade's surface through specially shaped holes in such a way that it forms a film of air over the blade to insulate the surface from the hot gases.

Air used for turbine system cooling is bled from one of the higher stages of the compressor, and while its temperature is greater than 1,000 °F (about 537.76 °C), it is far cooler than the gases that drive the turbine. The air flows through the hollow blades and exits with the exhaust gases. It is necessary to cool only the turbine inlet guide vanes and the first stage turbine blades. The gases lose enough of their energy when passing the first stage that their temperature drops to within the range allowed for succeeding stages.

Words & Expressions

1. expose   揭发, 使暴露
2. absorption   吸收
3. nozzle diaphragm   喷管挡栅板

课文翻译

4. deflection 偏移
5. disproportionately 不相称,不成比例地
6. impulse 冲击
7. cartridge 筒
8. bearing 轴承
9. self-aligning 自动对准
10. aerofoil 翼型
11. hollow 孔洞的,中空的
12. flange 法兰
13. bolt 螺栓,用螺栓固定
14. perimeter 周长;周围,边界
15. attachment 连接
16. throat 咽喉,颈前部
17. serration 锯齿状,锯齿状突起
18. stiffen (使)变硬,(使)强硬
19. leakage 泄露
20. shroud 保护罩
21. abradable 耐磨的
22. clearance 间隙
23. shroudless 无罩的
24. creep 蠕变
25. permanent 永久的
26. fatigue 疲劳
27. insulate 隔绝,隔热

Exercise

Ⅰ. Fill in the following blanks according to the text.

1. The turbine has the task of providing the power to drive the compressor and _____ and, in the case of engines which do not make use solely of a jet for propulsion, of providing shaft power for a propeller or rotor.

2. For a given output, the gas velocities, _____, and hence losses, are reduced in proportion to the square of mean blade speeds.

3. The nozzle guide vanes are of an _____ shape with the passage between adjacent vanes forming a convergent duct.

4. The nozzle guide vanes are usually of _____ form and may be cooled by passing compressor delivery air through them to reduce the effects of high thermal stresses and gas loads.

5. The shroud is made up by a small segment at the tip of each blade which forms a peripheral ring around the blade tips. An _____ lining in the casing may also be used to reduce gas leakage.

Ⅱ. Translate the following sentences into Chinese.

1. The exact amount of energy absorption by the turbine is determined by the load the turbine is driving (i.e., the size and type of the compressor, the number of accessories, and the load applied by the other turbine stages).

2. Stress in the turbine disc increases as the square of the speed, therefore to maintain the same stress level at higher speed, the sectional thickness, hence the weight, must be increased disproportionately. For this reason, the final design is a compromise between efficiency and weight.

3. The design of the nozzle guide vane and turbine blade passages is based broadly on aerodynamic considerations, and to obtain optimum efficiency, compatible with compressor and combustion design, the nozzle guide vanes and turbine blades are of a basic aerofoil shape.

4. The turbine blades are of an aerofoil shape, designed to provide passages between adjacent blades that give a steady acceleration of the flow up to the "throat", where the area is smallest and the velocity reaches that required at the exit to produce the required reaction.

5. Air used for turbine system cooling is bled from one of the higher stages of the compressor, and while its temperature is greater than 1,000 °F, it is far cooler than the gases that drive the turbine. The air flows through the hollow blades and exits with the exhaust gases.

Lesson 7   Exhaust Section

课文音频

Aero gas turbine engines have an exhaust system which passes the turbine discharge gases to atmosphere at a velocity, and in the required direction, to provide the resultant thrust. The velocity and pressure of the exhaust gases create the thrust in the turbo-jet engine but in the turbo-propeller engine only a small amount of thrust is contributed by the exhaust gases, because most of the energy has been absorbed by the turbine for driving the propeller. The design of the exhaust system therefore, exerts a considerable influence on the performance of the engine. The areas of the jet pipe and propelling or outlet nozzle affect the turbine entry temperature, the mass airflow, and the velocity and pressure of the exhaust jet.

The temperature of the gas entering the exhaust system is between 550 °C and 850 °C according to the type of engine and with the use of afterburning can be 1,500 °C or higher. Therefore, it is necessary to use materials and a form of construction that will resist distortion and cracking, and prevent heat conduction to the aircraft structure.

A basic exhaust system is shown in Figure 7.1. The use of a thrust reverser, a noise suppressor and a two-position propelling nozzle entails a more complicated system as shown in Figure 7.2. The low bypass engine may also include a mixer unit to encourage a thorough mixing of the hot and cold gas streams.

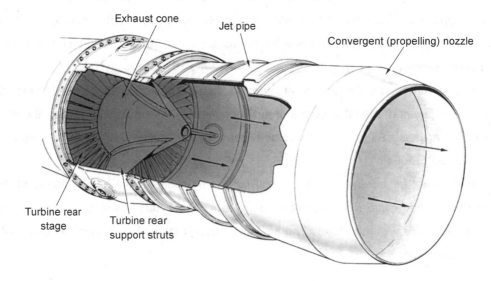

Figure 7.1 A basic exhaust system

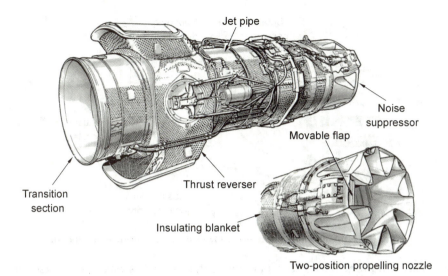

Figure 7.2   An exhaust system with thrust reverser, noise suppressor and two-position propelling nozzle

7.1   Exhaust Nozzles

The rear opening of a turbine engine exhaust duct is called the exhaust nozzle. The nozzle acts as an orifice whose size determines the density and velocity of exiting gases. Nozzle shapes are either convergent or convergent-divergent.

Convergent exhaust nozzle

Most aircraft that produce exhaust gas at subsonic velocities use a convergent exhaust nozzle to increase exhaust gas velocities.

Adjusting the area of an exhaust nozzle changes both engine performance and exhaust gas temperature. Some engines are trimmed to their correct exhaust gas temperature by altering the exhaust nozzle area. To do this, you can bend small tabs to change the area or fasten small adjustable pieces, called trim, around the perimeter of the nozzle.

Convergent-divergent nozzle

Whenever the engine pressure ratio is high enough to produce supersonic exhaust gas velocities, a convergent-divergent nozzle will produce more thrust. The advantage of a convergent-divergent nozzle is greatest at high Mach numbers because of the resulting higher pressure ratio across the engine exhaust nozzle. Figure 7.3 shows the principle of a convergent-divergent nozzle.

To ensure that a constant weight (or volume) of gas flows past any given point at

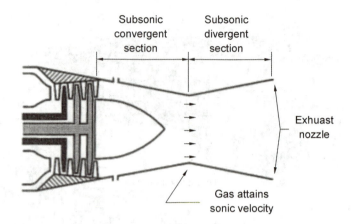

Figure 7.3　A diagram of the principle of a convergent-divergent nozzle

sonic velocity, the rear part of a supersonic exhaust duct is enlarged to accommodate the additional weight or volume of a gas flowing at supersonic velocities. Otherwise, the nozzle would not operate efficiently.

7.2　Reciprocating Engine Exhaust Systems

The reciprocating engine exhaust system is fundamentally a scavenging system that collects and disposes of the high temperature, noxious gases being discharged by the engine. Its main function is to efficiently dispose of the gases while ensuring complete safety of the airframe and the occupants of the aircraft. The exhaust system can perform many useful functions, but its first duty is to provide protection against the potentially destructive action of the exhaust gases. Modern exhaust systems, though comparatively light, adequately resist high temperatures, corrosion, and vibration to provide long, trouble-free operation with minimum maintenance.

There are two general types of exhaust systems in use on reciprocating aero engines: the short stack (open) system and the collector system. The short stack system is generally used on non-supercharged engines and low-powered engines where noise level is not too objectionable. The collector system is used on most large non-supercharged engines and all turbosupercharged engines and installations on which it would improve nacelle streamlining or provide easier maintenance in the nacelle area. On turbosupercharged engines, the exhaust gases must be collected to drive the turbine compressor of the supercharger. Such systems have individual exhaust headers that have a common outlet. From this outlet, the hot exhaust gas is routed via a tailpipe to the turbosupercharger that drives the turbine. Although the collector system raises the back pressure of the exhaust system, the gain in horsepower from turbosupercharging more than offsets the loss in horsepower that results from increased back pressure. The short stack system is relatively simple, and its removal and installation consists essentially of removing and installing the hold-down nuts and clamps. Short stack systems have limited use on most modern aircraft.

In Figure 7.4, the location of typical collector exhaust system components of a horizontally opposed engine is shown in a side view. The exhaust system in this installation consists of a down-stack from each cylinder, an exhaust collector tube on each side of the engine, and an exhaust ejector assembly protruding aft and down from each side of the firewall. The down-stacks are connected to the cylinders with high temperature locknuts and secured to the exhaust collector tube by ring clamps. A cabin heater exhaust shroud is installed around each collector tube.

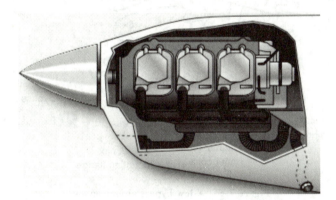

Figure 7.4 Location of a typical collector exhaust system

7.3 Turbofan Engine Exhaust Systems

In a low bypass ratio engine, the two flows are combined by a mixer unit (Figure 7.5) which allows the bypass air to flow into the turbine exhaust gas flow in a manner that ensures thorough mixing of the two streams.

In high bypass ratio engines, the two streams are usually exhausted separately. The hot and cold nozzles are co-axial and the area of each nozzle is designed to obtain maximum efficiency. However, an improvement can be made by combining the two gas flows within a common, or integrated, nozzle assembly. This assembly partially mixes the gas flows prior to ejection to atmosphere. An example of both types of high bypass exhaust system is shown in Figure 7.6.

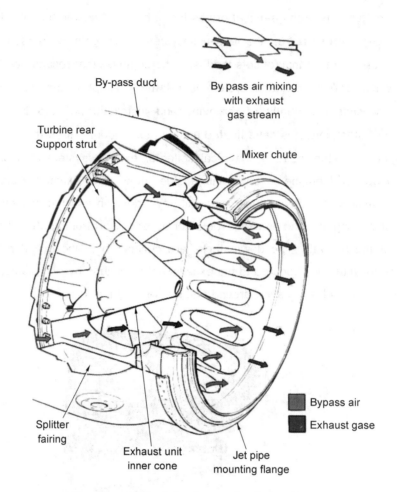

Figure 7.5　A low bypass air mixer unit

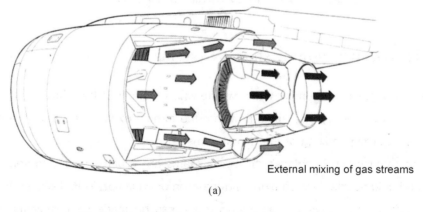

(a)

Figure 7.6　High bypass ratio engine exhaust systems

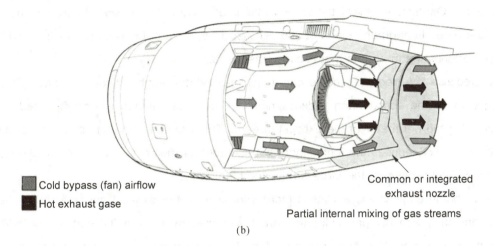

Cold bypass (fan) airflow
Hot exhaust gase

Common or integrated exhaust nozzle
Partial internal mixing of gas streams

(b)

Figure 7.6(Continued)

7.4 Noise Suppressors

Because most major airports are located near large cities, the need to minimize turbine exhaust noise is obvious. Aircraft and engine manufacturers have worked with regulators and operators to improve noise reduction technologies and techniques with each generation of products. When possible, new technologies and techniques are retrofitted to earlier designs to reduce noise.

Noise suppressors used on the ground include portable devices positioned near the rear of an engine during periods of prolonged ground operation. Most large airports have blast fences and designated run-up areas, and aircraft operations are restricted to certain times of the day.

Older turbojet engines produce a combination of noise frequencies at very high levels. Although a turbojet compressor produces a great deal of high frequency sound, this noise decreases rapidly as the distance from the source increases. Turbojet exhaust also produces noise at a wide range of frequencies and at very high energy levels. This noise is damaging to human hearing over great distances. One solution to turbojet exhaust noise is the use of a corrugated perimeter noise suppressor that helps break up the exhaust flow and raises its noise frequency. Some older engines can be fitted with hush kits that reduce their noise emissions.

Newer engines use a variety of techniques to reduce harmful noise. For example, some turbofan engines blend fan discharge air with the exhaust gases to reduce sound emission. On these engines, the sound at the inlet is likely to be louder than sound from the tail pipe. In addition, the inlet and exhaust ducts on turbofan engines are lined with sound-attenuating materials that reduce noise levels.

Because low frequency noise can linger at a relative high volume, increasing the frequency of the sound often reduces noise. Frequency change is accomplished by increasing the perimeter of the exhaust stream. This provides more space for cold and hot air to mix, reduces the tendency of hot and cold air molecules to shear against one another, and breaks up the large turbulence in jet wake.

Turbofan engines extract much more energy from the exhaust gases to drive the fan, and their exhaust gas velocities are lower than those of a turbojet engine with comparable power. For these reasons, turbofan engines do not produce enough noise to require noise suppressors. For turbofan engines, the noise reduction locations include the inner wall of the inlet cowl, the inner wall of the fan case, and the inner wall of the tail nozzle.

 Words & Expressions

课文翻译

1.	exhaust system	排气系统
2.	propeller(=propellor)	螺旋桨,推进器
3.	afterburning	二次燃烧,后燃
4.	crack	破裂,变形
5.	suppressor	干扰抑制器
6.	exhaust nozzle	喷管
7.	orifice	孔,洞口
8.	trim	修整,完善;配平
9.	volume	体积
10.	scavenge	清除污物
11.	noxious	有害的,有毒的
12.	occupant	占有人,居住者,乘客
13.	non-supercharged	非增压的
14.	nacelle	飞机的引擎机舱
15.	turbosupercharged	装有涡轮增压器的
16.	tailpipe	排气管
17.	hold-down nut	压紧螺母

Module Two Structure of Aero Engine | 67

18.	clamp	夹具
19.	mixer	混合器
20.	airport	飞机场
21.	regulator	监管机构
22.	operator	运营商
23.	retrofit [ˈretrəʊfɪt]	翻新,改型
24.	blast	爆炸
25.	fence	围栏
26.	corrugated	波纹的
27.	hush	安静
28.	attenuate	(使)变小,减弱
29.	linger	徘徊
30.	molecule	分子
31.	extract	获得,提取
32.	cowl	整流罩
33.	opposed engine	对置式发动机

 Exercise

Ⅰ. Fill in the following blanks according to the text.

1. The use of a thrust reverser, a noise _____ and a two-position propelling nozzle entails a more complicated system as shown in Figure 7.2.

2. The nozzle acts as an _____ whose size determines the density and velocity of exiting gases.

3. The reciprocating engine exhaust system is fundamentally a _____ system that collects and disposes of the high temperature, noxious gases being discharged by the engine.

4. The collector system is used on most large non-supercharged engines and all turbosupercharged engines and installations on which it would improve _____ streamlining or provide easier maintenance in the nacelle area.

5. One solution to turbojet exhaust noise is the use of a _____ perimeter noise suppressor that helps break up the exhaust flow and raises its noise frequency. Some older engines can be fitted with hush kits that reduce their noise emissions.

6. For turbofan engines, the noise reduction locations include the inner wall of the

inlet _____, the inner wall of the fan case, and the inner wall of the tail nozzle.

II. Translate the following sentences into Chinese.

1. The velocity and pressure of the exhaust gases create the thrust in the turbo-jet engine but in the turbo-propeller engine only a small amount of thrust is contributed by the exhaust gases, because most of the energy has been absorbed by the turbine for driving the propeller.

2. To ensure that a constant weight (or volume) of gas flows past any given point at sonic velocity, the rear part of a supersonic exhaust duct is enlarged to accommodate the additional weight or volume of a gas flowing at supersonic velocities. Otherwise, the nozzle would not operate efficiently.

3. Although the collector system raises the back pressure of the exhaust system, the gain in horsepower from turbosupercharging more than offsets the loss in horsepower that results from increased back pressure. The short stack system is relatively simple, and its removal and installation consists essentially of removing and installing the hold-down nuts and clamps.

4. Aircraft and engine manufacturers have worked with regulators and operators to improve noise reduction technologies and techniques with each generation of products. When possible, new technologies and techniques are retrofitted to earlier designs to reduce noise.

5. On these engines, the sound at the inlet is likely to be louder than sound from the tail pipe. In addition, the inlet and exhaust ducts on turbofan engines are lined with sound-attenuating materials that reduce noise levels.

Module Three Engine System

 ## Lesson 8 Fuel System

课文音频

The aircraft's propulsion needs change depending on the stage of flight. During this process, aero engines must operate at various stages and consume various amounts of fuel. Providing clean, de-aerated, pressurized, and metered gasoline to the engine in various operating conditions to suit its various demands at these stages is one of the engine fuel system's primary functions. The fuel pressure system, fuel control system, and fuel indicator system are the three fully functional subsystems that make up the engine fuel system. These technologies provide effective and flexible fuel supply during the flight, adjusting to the engine's fluctuating demands.

8.1 Turbine engine fuel

The end of World War II marked the highest demand for aviation gasoline (AVGAS). However, the demand for aviation gasoline started to drop with the introduction of aero turbine engines, and it was replaced by jet fuel. Jet fuel is characterized by the following:

First, it has a low enough freezing point. Freezing point is the temperature at which a liquid becomes a solid. If the freezing point of the fuel is not low enough, it is easy to cause the production of ice crystals inside the fuel, resulting in the reduction of the fuel flow rate, which affect the normal oil supply of the engine. The freezing point of jet fuel must be below $-40°C$.

Second, it has a suitable flash point. The flash point is the temperature at which a mixture of gas formed by fuel and air from outside burns immediately upon contact with a flame.

The flash point is too low, the stability of the fuel is worse, and the aircraft is prone to fire. The flash point is too high, fuel ignition is difficult, it is easy to cause the engine flame out.

Third, it has the appropriate volatility. Jet fuel, like other liquids, is vaporized by a reduction in atmosphere pressure. The higher the flight altitude, the lower the atmosphere pressure, and the more likely it is to lead to fuel vaporization. Higher fuel vaporization rate is favorable in cold environments, as well as during engine cold starts, however, excessive vaporization will cause increased fuel loss during the vaporization process. Therefore, jet fuel needs to have suitable vaporization properties.

Fourth, it has good lubrication performance. In addition to combustion, fuel is used to lubricate moving parts in the fuel system (e. g. oil pump). In order to ensure the normal operation of the fuel system, the lubrication performance of the fuel is particularly important.

Fifth, it has a poor water absorption performance. Jet fuel must not be easy to maintain moisture to reduce fuel pollution. Fuel tank corrosion can occur if water remains within the tank, where it will cultivate bacteria.

8.2 Fuel pressure system

There are two types of fuel pressure systems: low pressure and high pressure. To guarantee smooth functioning, fuel must be supplied to the engine high pressure pump by the low pressure system. The fuel pressure inside the combustion chamber is raised to a level where the fuel is effectively atomized by the high pressure system. The fuel system's low pressure section consists of the fuel pressure sensor, flow sensor, temperature sensor, oil filter, fuel heater, and air heater, in addition to the low pressure pump. The components of a high pressure system are the fuel controller, fuel nozzle, and high pressure pump. The primary component, such as the fuel controller, regulates the oil supply and keeps the engine safe while it is operating.

Pump

Pumps fall into two major groups:

(1) Positive displacement pumps which are divided into reciprocating pumps and rotary pumps;

(2) Rotodynamic pumps which are divided into centrifugal pumps and axial flow pumps.

Depending on flow rate demand, low flow rate pumps can be used known as gear pumps and vane pumps. They are both rotary pumps. For high flow rate applications, in-line piston pumps and Strato power pumps are used. They are both from the family of reciprocating pumps.

Gear pump

A gear pump is a low flow rate pump that incorporates externally meshing gears enclosed in housing. This is a more efficient pump than a Gerotor applicable for the pressure less than 1,000 psi(1 psi=6,894.76 Pa). The fluid pressure limit in a gear pump is 1,500 psi. The fluid is trapped between the gear teeth and the housing wall, and is transferred from the inlet to the outlet port as it is carried around between the teeth and the housing wall. Figure 8.1 shows a gear pump.

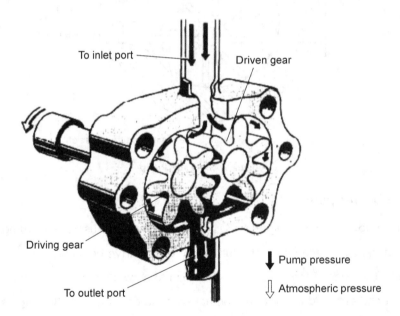

Figure 8.1 Gear pump

Vane pump

Vane pumps are used where higher flow rates are required. The fluid pressure is up to 2000 psi. In a vane pump, the rotor is slotted and each slot is fitted with a rectangular vane(Figure 8.2). These vanes have some latitude to move outward in their respective slots. The rotor and vanes are enclosed in a housing, the inner surface of which is offset with the drive axis in unbalanced vane pumps. As the rotor turns, centrifugal force keeps the vanes in contact with the wall of the housing. The vanes divide the area between the rotor and housing into a series of chambers. These chambers vary in size according to their respective positions around the shaft. The fluid inlet port is located in the part of the pump where the chambers are expanding in size so that the partial vacuum, or low-pressure area formed by this expansion allows liquid to flow into the pump. The liquid is trapped between the vanes and carried to the fluid outlet side of the pump. The chambers contract in size on the outlet side, and this action forces the liquid through the outlet port.

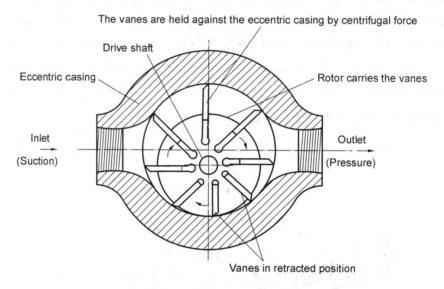

Figure 8.2 Vane Pump

Strato power pump

In Strato power pumps(Figure 8.3), piston motion is caused by rotating the drive cam, thus displacing each piston the full height of the drive cam during each revolution of the shaft. The ends of the pistons are attached to a swash plate supported by a freed centre pivot and are held in constant contact with the cam face. As the high side of the rotating drive cam depresses one side of the swash plate, the other side of the swash plate is withdrawn an equal amount, thus moving the pistons with it. Two creep plates are provided to decrease wear on the revolving cam.

Filter

The purpose of an oil filter is to keep oil contaminants out of the engine fuel system, where they can lead to wear on components and oil passage blockages. As a result, an oil filter with fine mesh or coarse mesh will be installed in the fuel system. Fine oil filters(fine mesh), also known as low pressure oil filters, are typically located near the engine's fuel system's starting location to keep foreign object debris (FOD) from getting inside. High pressure oil filters, also known as coarse oil filters, provide a

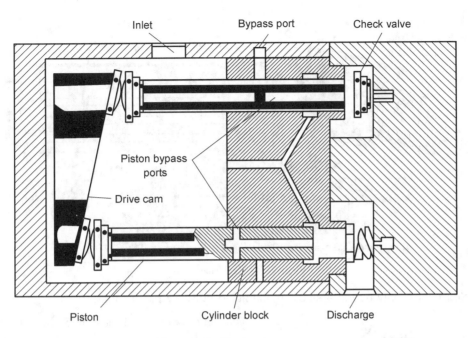

Figure 8.3 Strato power pump

protective role by preventing fine oil filters downstream from being damaged by nozzle blockages. Coarse oil filters are often placed in a position before fuel enters the nozzle. Typically, fine oil filters are disposable, require routine replacement, or replacement after clogging, and come with a bypass valve and a blockage indicator mechanism. The coarse oil filter uses metal mesh so that it can be cleaned and reused using ultrasonic technology.

Fuel heater

The moisture exists in fuel could freeze while the turbine engine operates at a high altitude. As a result, there could be blockage in the oil filter. The solution to this issue is a fuel heater. Air/fuel heaters (Figure 8.4) or oil/fuel heaters are the primary fuel heaters used in modern aero-engines. The oil / fuel heater is typically put on fuel pipelines with low pressure. The compressor's hot air is used by the air/fuel heater to heat the fuel. Through a switch located in the cockpit, the pilot controls the fuel heater. The flight crew can determine when to heat fuel by using the fuel temperature gauge and filter bypass indication light information.

Fuel spray nozzle

Different engine types employ different fuel nozzle designs, the two most popular types are the simplex and duplex. The duplex nozzle needs a dual manifold and a pressurizing valve or flow divider to divide primary and secondary fuel flow, but the

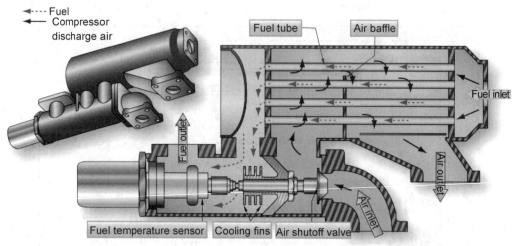

Figure 8.4 Air/fuel heater

simplex nozzle only needs a single manifold to delivery proper fuel.

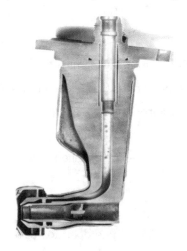

Figure 8.5 Simplex nozzle

The simplex nozzle(Figure 8.5)was the first nozzle type used in turbine engine and replaced in most installations with the duplex nozzle, which gave better atomization at starting and idling speed. The simplex nozzle is still being used in several installation. Each of the simplex nozzles consists of a nozzle tip, an insert, and a strainer made up of a fine mesh screen and a support.

The duplex fuel nozzle(Figure 8.6)is widely used in present day gas turbine. As mentioned previously, its use requires a flow divider, but at the same time it offers a desirable spray pattern for combustion over a wide range of operating pressure.

Flow divider

A flow divider(Figure 8.7)creates primary and secondary fuel supplies that are discharged through separate manifolds, providing two separate fuel flows. Fuel from the fuel control enters the inlet of the flow divider and passes through an orifice and then on to the primary nozzles. A passage in the flow divider directs the fuel flow from both sides of the orifice to a chamber. This chamber contains a differential pressure bellows, a viscosity compensated restrictor (VCR), and a surge dampener. During engine start, fuel pressure is applied to the inlet port and across the VCR, surge

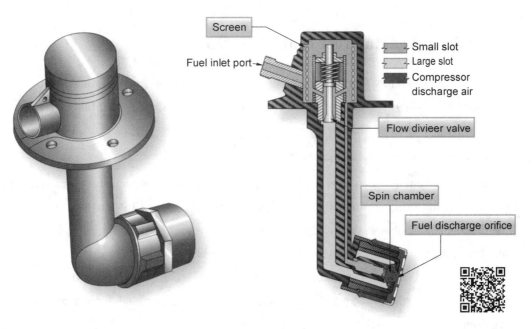

Figure 8.6 Duplex fuel nozzle

dampener, and on to the primary side of the nozzles. Fuel is also applied under pressure to the outside of the flow divider bellows and through the surge dampener to the inside of the flow divider bellows. This unequal pressure causes the flow divider valve to remain closed. When fuel flow increases, the differential pressure on the bellows also increases. At a predetermined pressure, the bellows compresses, allowing the flow divider valve to open. This action starts fuel flow to the secondary manifold, which increases the fuel flow to the engine. This fuel flows out of the secondary opening in the nozzles.

Figure 8.7 A flow divider

8.3 Turbine engine fuel control system

Turbine engine fuel control system can be divided into three basic groups: hydromechanical, hydromechanical/electronic, and full authority digital engine (or electronics) control (FADEC) systems.

Hydromechanical fuel control system

A pure hydromechanical fuel control system has no electronic interface assisting in computing or metering the fuel flow. It is also generally driven by the gas generator gear train of the engine to sense engine speed. Other mechanical engine parameters that are sensed are compressor discharge pressure, burner pressure, exhaust temperature, and inlet air temperature and pressure. Once the computing section determines the correct amount of fuel flow, the metering section delivers the fuel to the engine fuel system through cams and servo valves.

Hydromechanical/electronic fuel control system

Hydromechanical/electronic fuel control involves the addition of a remotely located electronic engine control (EEC) to the hydromechanical fuel control system to manage fuel flow. The turbine engine fuel system's main functions include pressurizing fuel, monitoring the fuel flow, and supplying fuel to the engine's combustion chamber. Fuel flow is controlled by a hydromechanical fuel assembly, which contains a fuel shutoff section and a metering section. There are manual and automatic modes available for the system. Fuel metering in automatic mode is controlled by the EEC. Hydromechanical control system takes over when operating in manual mode.

FADEC fuel control system

A FADEC system has been developed to control fuel flow on most new turbine engine models. A true FADEC system has no hydromechanical fuel control backup system. The system uses electronic sensors that feed engine parameter information into the EEC. The EEC gathers the needed information to determine the amount of fuel flow and transmits it to a fuel metering valve. The fuel metering valve simply reacts to the commands from the EEC. The EEC is a computer that is the computing section of the fuel delivery system and the fuel metering valve meters the fuel flow. FADEC systems are used on many types of turbine engines from APUs to the largest propulsion engines.

Module Three　Engine System ｜ 77

 Words & Expressions

课文翻译

1. de-aerated　　　　　　　　　　　去除蒸汽的
2. gasoline　　　　　　　　　　　　汽油
3. freezing point　　　　　　　　　 冰点
4. flash point　　　　　　　　　　　闪点
5. volatility　　　　　　　　　　　　挥发性
6. lubrication performance　　　　　润滑性能
7. water absorption performance　　吸水性能
8. moisture　　　　　　　　　　　　湿润
9. pollution　　　　　　　　　　　　污染
10. tank　　　　　　　　　　　　　　燃料箱
11. bacteria　　　　　　　　　　　　细菌
12. atomization　　　　　　　　　　 雾化
13. rotodynamic pump　　　　　　　回转动力式泵
14. Strato power pump　　　　　　　斯特拉托动力泵
15. gear pump　　　　　　　　　　　齿轮泵
16. psi(pound per square inch)　　　磅每平方英寸（压强单位）
17. vane pump　　　　　　　　　　　叶片泵
18. housing　　　　　　　　　　　　腔，室
19. centrifugal force　　　　　　　　离心力
20. drive cam　　　　　　　　　　　 驱动凸轮
21. swash plate　　　　　　　　　　 旋转斜盘，倾斜盘
22. creep plate　　　　　　　　　　　蠕盘
23. blockage　　　　　　　　　　　　堵塞
24. fine mesh　　　　　　　　　　　 细滤网
25. fuel heater　　　　　　　　　　　燃油加热器
26. stainer　　　　　　　　　　　　　油滤
27. flow divider　　　　　　　　　　 分流器
28. secondary manifold　　　　　　 二级歧管
29. hydromechanical fuel control　　液压机械式燃油控制
30. hydromechanical/electronic fuel control　　液压机械/电子式燃油控制
31. FADEC(full authority digital electronics control)　　全权限数字电子控制

Exercise

Ⅰ. Fill in the following blanks according to the text.

1. Providing clean, _____, pressurized, and metered _____ to the engine in various operating conditions to suit its various demands at these stages is one of the engine fuel system's primary functions.

2. _____ is the temperature at which a liquid becomes a solid.

3. The _____ is the temperature at which a mixture of gas formed by fuel and air from outside burns immediately upon contact with a flame.

4. In addition to _____, fuel is used to lubricate moving parts in the fuel system (e. g. oil pump).

5. Jet fuel must not be easy to maintain _____ to reduce fuel _____. Fuel tank corrosion can occur if water remains within the _____ where it will cultivate _____.

Ⅱ. Translate the following sentences into Chinese.

1. The fuel pressure system, fuel control system, and fuel indicator system are the three fully functional subsystems that make up the engine fuel system.

2. If the freezing point of the fuel is not low enough, it is easy to cause the production of ice crystals inside the fuel , resulting in the reduction of the fuel flow rate, which affect the normal oil supply of the engine.

3. The flash point is too low, the stability of the fuel is worse, and the aircraft is prone to fire.

4. Fuel tank corrosion can occur if water remains within the tank where it will cultivate bacteria.

5. The fuel pressure inside the combustion chamber is raised to a level where the fuel is effectively atomized by the high pressure system.

Lesson 9　Lubrication and Cooling Systems

课文音频

When the turbine engine is working, the contact surfaces of each component make a relative motion at a very high speed. Although the contact surface of each part looks very smooth, it still shows a certain roughness under the microscope. In order to prevent the relative motion and dry friction, the two pieces' relative motion will be limited by the rough protrusion of the surfaces striking with one another. Lubrication is to form a layer of scavenge oil film on the contacting metal surfaces, so that the scavenge oil fills the uneven surfaces of the parts, and separates the contacting parts by the oil film, so that the dry friction between the surfaces of the relatively moving parts turns into liquid friction, thus greatly reducing the friction resistance.

9.1 Scavenge oil

The function of scavenge oil

The primary function of the oil system is to deliver clean oil, at the proper temperature and pressure, to the lubrication position in order to guarantee the engine runs as normally as possible. The oil system's primary purposes are anticorrosion, cleaning, cooling, and lubrication.

Lubrication: By applying a layer of oil film to the surfaces of the moving parts that are in touch with each other, but not directly, the friction between the metal surface and the oil film is reduced, allowing for mutual movement.

Cooling: Reduces temperature and takes away heat; the oil absorbs heat from bearings and other hot parts, and transfers heat to the fuel or air for cooling.

Anticorrosion: In order to avoid oxidation and corrosion, surfaces of metal pieces are seperated from the air by covering them with an oil film of a specific thickness.

The kind of scavenge oil

There are two types of engine oils, namely synthetic oil and petroleum oil, which are commonly used in reciprocating engines. Different engine manufacturers will impose relevant limits on the types of oil used in their own engine products. As a result, it is essential to select oil in accordance with the manufacturer's recommendations.

Performance metric

The terms viscosity, viscosity index, cloud point, pour point, flash point, and fire point are used to describe the scavenge oil performance metrics.

Viscosity: The relative movement between oil layers, as well as the amount of the friction force between the oil molecules, reflect the oil's fluidity; a high viscosity indicates poor fluidity, while a low viscosity indicates good fluidity.

Flash point: When the oil is heated, oil and gas develop on the surface. When heated to a specific temperature, the oil and gas spreading on the scavenge oil surface quickly produce a quick spark when they come into contact with an open flame, but they cannot maintain their combustion temperature. This specific temperature is the flash point of the scavenge oil.

Cloud point: The temperature at which the oil begins to become cloudy. That is, at that temperature, the wax content, normally held in solution, begins to solidify and separate into tiny crystals, causing the oil to appear cloudy or hazy.

Pour point: The lowest temperature at which the scavenge oil can flow or can be poured.

9.2 Turbine lubrication system

The turbine lubrication system consists of dry-sump and wet-sump lubrication systems. The wet-sump lubrication system stores the lubrication oil in the engine proper, while the dry-sump lubrication system utilizes an external tank mounted on the engine or somewhere in the aircraft structure near the engine.

Dry-sump lubrication system

The turbine dry-sump lubrication system is representative of turbine engines using a dry-sump system (Figure 9.1). This lubrication system is a pressure regulated, high-pressure design. It consists of the pressure, scavenge, and breather subsystems.

The pressure system supplies oil to the main engine bearing and accessory drives. The scavenger system returns the oil to the engine oil tank that is usually mounted on the compressor case. It is connected to the inlet side of the pressure oil pump and completes the oil flow cycle. A breather system connecting the individual bearing compartments and the oil tank with the breather pressurizing valve completes the engine lubrication system. In a dry-sump lubrication system, the oil supply is carried in a tank mounted on the engine. With this type of system, a larger oil supply can be carried, and the temperature of the oil can be readily controlled.

The dry-sump variable-pressure lubrication system (Figure 9.2) uses the same basic subsystems that the regulated systems used (pressure scavenge breather). The main difference is that the pressure in this system is not regulated by a regulating bypass valve. Most large turbofan engine pressure systems are variable-pressure

Module Three Engine System | 81

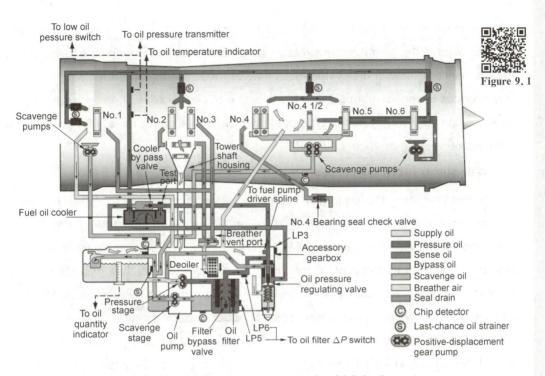

Figure 9.1 Typical dry-sump pressure regulated lubrication system

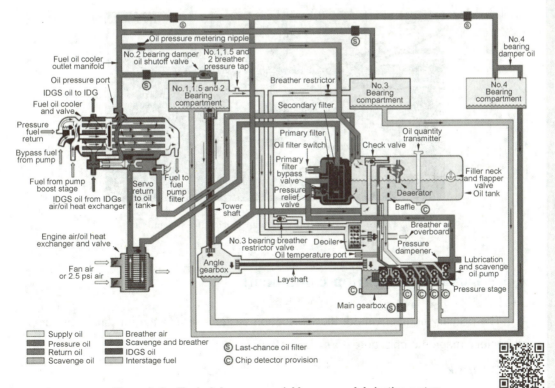

Figure 9.2 Typical dry-sump variable-pressure lubrication system

systems in which the pump outlet pressure (oil pressure) depends on the engine rpm. In other words, the pump output pressure is proportional to the engine speed. Since the resistance to flow in the system does not vary much during operation and the pump only changes in speed, the pressure is a function of engine speed. As an example, oil pressure can vary widely in this type of system, from 100 psi to over 260 psi, with the relief valve opening at about 540 psi.

Wet-sump lubrication system

In some engines, the lubrication system is the wet-sump type. There are relatively few engines using a wet-sump lubrication system. A schematic diagram of a wet-sump lubrication system is shown in Figure 9.3. The components of a wet-sump lubrication system are similar to those of a dry-sump lubrication system. The major difference between the two systems is the location of the oil reservoir. The reservoir for the wet-sump lubrication system may be the accessory gear case or it may be a sump mounted on the bottom of the accessory case. Regardless of configuration, reservoirs for wet-sump lubrication systems are an integral part of the engine and contain the bulk of the engine oil supplied.

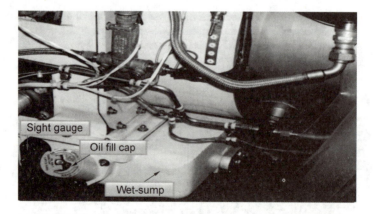

Figure 9.3 Typical wet-sump lubrication system

9.3 Turbine lubrication component

The primary components of the lubrication system include the oil tank, oil pump, oil cooler, magnetic chip detectors, and others.

Oil tank

Gas turbine engines frequently have a separate oil tank attached to an easily accessible area of the engine casing, allowing route maintenance staff to execute their job. The tank capacity is governed by the engine's demand, which is primarily influenced by three factors: the oil consumption of the engine, the thermal expansion of the oil after it has been combined with air, and the need for safe storage. A view of a typical oil tank is shown in Figure 9.4. It is designed to furnish a constant supply of oil to the engine during any aircraft attitude. This is done by a swivel outlet assembly mounted inside the tank, a horizontal baffle mounted in the center of the tank, two flapper check valves mounted on the baffle, and a positive vent system. The swivel outlet fitting is controlled by a weighted end that is free to swing below the baffle. The flapper valves in the baffle are normally open; they close only when the oil in the bottom of the tank tends to rush to the top of the tank during decelerations. This traps the oil in the bottom of the tank where it is picked up by the swivel fitting. A sump drain is located in the bottom of the tank. The vent system inside the tank is so arranged that the airspace is vented at all times even though oil may be forced to the top of the tank by deceleration of the aircraft. All oil tanks are provided with expansion space. This allows expansion of the oil after heat is absorbed from the bearings and gears and after the oil foams as a result of circulating through the system.

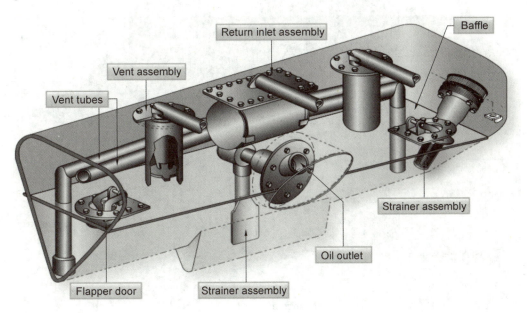

Figure 9.4 Oil tank

Oil pump

The oil pump's work is to cycle the oil inside the engine. The pump that transports the oil from the oil tank to the bearing chamber and gear box is known as the oil supply pump, and it is equipped with an oil supply valve to avoid leakage and damage to certain elements of the sliding system caused by the system's high pressure. The return pump collects lubricated oil and returns it to the oil tank.

The pumps may be one of several types, each type having certain advantages and limitations. The two most common oil pumps are the gear-type and gerotor-type, with the gear-type being the most commonly used. Each of these pumps has several possible configurations. The gear-type oil pump has only two elements: one for pressurizing oil and one for scavenging oil (Figure 9.5). However, some types of pumps may have several elements: one or more elements for pressurizing and two or more for scavenging oil. The clearances between the gear teeth and the sides of the pump wall and plate are critical to maintain the correct output of the pump.

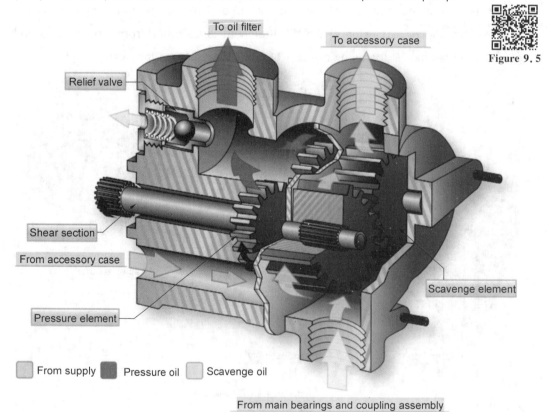

Figure 9.5 Oil pump

The gerotor oil pump, like the gear oil pump, usually contains a single element for pressurizing oil and several elements for scavenging oil. A typical set of gerotor pumping elements is shown in Figure 9.6. Each set of gerotors is separated by a steel plate, making each set an individual pumping unit consisting of an inner and an outer element. The small star-shaped inner element has external lobes that fit inside and are matched with the outer element that has internal lobes. The small element fits on and is keyed to the pump shaft and acts as a drive for the outer free-turning element. The outer element fits within a steel plate having an eccentric bore. In one engine model, the oil pump has four elements: one for feeding oil and three for scavenging oil. In some other models, pumps have six elements: one for feeding oil and five for scavenging oil. In each case, the oil flows as long as the engine shaft is turning.

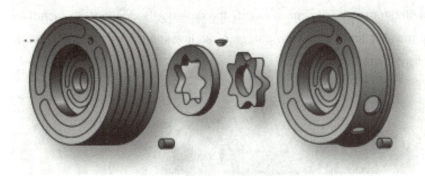

Figure 9.6 Typical gerotor pumping elements

Oil cooler

The role of oil cooler is to cool the oil to ensure the oil temperature within the limited operating range. The lubrication system of oil cooler installed on the scavenge oil road called the cold oil tank system. The oil cooler can also be separated into two types based on the cooling medium: air-oil cooler and fuel oil cooler. Fuel oil runs through honeycomb tubes, for heat exchange with the external flow oil. A baffler is installed to compel the oil to flow up and down for improved heat exchange.

When the temperature is low, the viscosity is high, or the pressure differential between the oil cooler inlet and outlet reaches a specific level, the bypass valve opens, and part of the oil runs through the oil cooler while the rest is supplied directly, ensuring a low temperature start.

Some engines' radiators contain an oil temperature sensor near the oil outlet that measures the temperature of the oil. Because the oil cooler is rather tiny in some small turbofan engines, the air-oil cooler is directly attached to the outer bypass valve, allowing the outer bypass air to travel directly through the oil cooler, resulting in oil cooling. When employing this layout, the oil bypass valve is typically located on the oil cooler. When the oil does not need to be cooled, the bypass valve opens and allows the oil to bypass the oil cooler.

Magnetic chip detector

Magnetic chip detectors (Figure 9.7) are used in the oil system to detect and catch ferrous (magnetic) particles present in the oil. Scavenge oil generally flows past chip detectors so any magnetic particles are attracted and stick to the chip detector. Chip detectors are placed in several locations but generally are in the scavenge lines for each scavenge pump, oil tank, and in the oil sumps. Some engines have one or more detectors. During maintenance, the chip detectors are removed from the engine and inspected for metal. If none is found, the detector is cleaned, replaced, and safety wired; if metal is found on a chip detector, an investigation should be made to find the source of the metal on the chip.

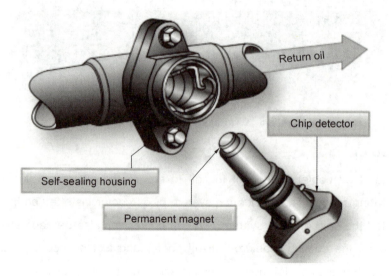

Figure 9.7　Magnetic chip detectors

9.4　Turbine cooling system

The gas turbine's combustion process is continuous, and the internal temperature rises to above 4,000 °F (about 2,204 °C) with providing an optimal air-fuel ratio of 15 : 1. However, in actuality, the engine can hold a significant amount of air in a percentage that exceeds the optimum. The engine's hot section was cooled to an acceptable temperature range of 1,500 − 2,100 °F (815.6 − 1148.9 °C) thanks to a large amount of surplus air.

The secondary air passing through the engine cools the combustion-chamber liners. The liners are constructed to induce a thin, fast-moving film of air over both the inner and outer surfaces of the liner. Can-annular-type burners frequently are provided with a center tube to lead cooling air into the center of the burner to promote high combustion efficiency and rapid dilution of the hot combustion gases while minimizing pressure losses. In all types of gas turbines, large amounts of relatively cool air join and mix with the burned gases aft of the burners to cool the hot gases just before they enter the turbines. Cooling-air inlets are frequently provided around the exterior of the engine to permit air to enter to cool the turbine case, the bearings, and the turbine nozzle. Internal air is bled from the engine compressor section and is vented to the bearings and other parts of the engine. Air vented into or from the engine is ejected into the exhaust stream. When located on the side of the engine, the case is cooled by outside air flowing around it. The engine exterior and the engine nacelle are cooled by passing fan air around the engine and the nacelle. The engine compartment frequently is divided into two sections. The forward section is referred to as the cold section and the aft section (turbine) is referred to as the hot section. Case drains drain almost potential leaks overboard to prevent fluids from building up in the nacelle.

Accessory zone cooling

Turbine powerplants can be divided into primary zones that are isolated from each other by fireproof bulkheads and seals. The zones are the fan case compartment, intermediate compressor case compartment, and the core engine compartment (Figure 9.8). Calibrated airflows are supplied to the zones to keep the temperatures around the engine at levels that are acceptable. The airflow provides for proper ventilation to prevent a buildup of any harmful vapors. Zone 1, for example, is around the fan case that contains the accessory case and the electronic engine control. The vapor in this area is vented by using ram air through an inlet in the nose cowl and is exhausted through a louvered vent in the right fan cowling.

If the pressure exceeds a certain limit, a pressure relief door opens and relieves the pressure. Zone 2 is cooled by fan air from the upper part of the fan duct and the vapor is exhausted at the lower end back into the fan air stream. This area has both fuel and oil lines, so removing any unwanted vapors would be important. Zone 3 is the area around the high-pressure compressor to the turbine cases. This zone also contains

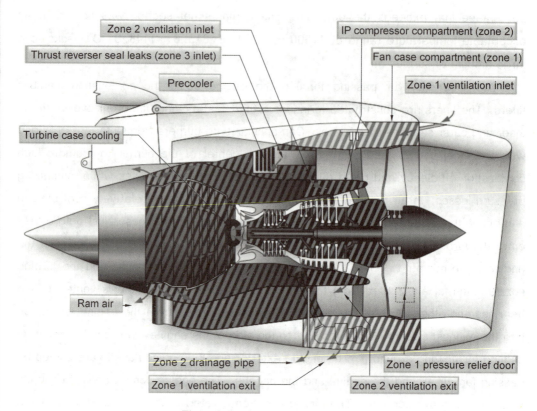

Figure 9.8 Accessory zone cooling

fuel and oil lines and other accessories. Air enters from the exhaust of the precooler and other areas and is exhausted from the zone through the aft edge of the thrust reverser inner wall and the turbine exhaust sleeve.

Words & Expressions

1. roughness 粗糙度
2. anticorrosion 防腐
3. synthetic oil 人工合成滑油
4. petroleum oil 从石油中提炼的滑油
5. cloud point 浊点，云点
6. tiny crystals 微小的晶体
7. breather subsystem 通气的子系统
8. accessory drive 附件驱动器
9. dry-sump variable pressure lubrication system 干式油底壳可变压力润滑系统

Module Three　Engine System | 89

10. turbine engine wet-sump lubrication system　　湿式油底壳润滑系统
11. bulk　　大部件
12. oil tank　　滑油箱
13. capacity　　容量
14. cycle　　使……循环
15. gerotor pump　　摆线泵
16. lobe　　叶轮
17. eccentric bore　　偏心孔
18. oil cooler　　滑油散热器
19. honeycomb tube　　蜂窝管
20. radiator　　散热器
21. magnetic chip detector　　磁屑探测器
22. thanks to　　由于
23. can-annular-type burner　　环管型火焰筒
24. exterior　　外部的
25. bulkhead　　舱壁，隔板，隔框
26. ventilation　　通风，通气
27. fan cowling　　风扇整流罩
28. thrust reverser　　推力反向器
29. exhaust sleeve　　排气套管
30. horizontal　　水平的

Exercise

Ⅰ. Fill in the following blanks according to the text.

1. Although the contact surface of each part looks very smooth, it still shows a certain _____ under the microscope.

2. The oil system's primary purposes are _____, cleaning, cooling, and lubrication.

3. There are two types of engine oils, namely _____ and _____, which are commonly used in reciprocating engines.

4. The pressure system supplies oil to the main engine _____ and accessory

drives.

5. The tank _____ is governed by the engine's demand, which is primarily influenced by three factors: the oil consumption of the engine, the thermal expansion of the oil after it has been combined with air, and the need for safe storage.

Ⅱ. Translate the following sentences into Chinese.

1. Although the contact surface of each part looks very smooth, it still shows a certain roughness under the microscope.

2. The oil system's primary purposes are anticorrosion, cleaning, cooling, and lubrication.

3. There are two types of engine oils, namely synthetic oil and petroleum oil, which are commonly used in reciprocating engines.

4. The pressure system supplies oil to the main engine bearing and accessory drives.

5. Gas turbine engines frequently have a separate oil tank attached to an easily accessible area of the engine casing, allowing route maintenance staff to execute their job.

 ## Lesson 10　Starting and Ignition Systems

课文音频

The starting and ignition systems are required for engine startup. The starting mechanism is utilized to move the engine from a static to a steady idling state, with the starter carrying the engine rotor through the accessory gearbox. The compressor rotates, drawing air into the engine. When the speed achieves self-acceleration state, the fuel system begins to provide oil, combining the air entering the combustion chamber with the fuel expelled from the nozzle to generate a fuel-air mixture. The fuel-air mixture burns when it is ignited by the ignition mechanism. The combustion produces a high temperature, and the high pressure bleed air drives the turbine's rotation. In this manner, the turbine and starter work together to continuously accelerate the compressor. The starter stops working when the speed reaches a predetermined point, and the turbine then turns the engine rotor to idle speed to finish the starting operation. The starting and ignition systems collaborate and work together during the start procedure.

10.1 Starting system

Gas turbine engine starter

Gas turbine engines are started by rotating the high-pressure compressor. On dual-spool, axial flow engines, the high pressure compressor and N1 turbine system is only rotated by the starter. To start a gas turbine engine, it is necessary to accelerate the compressor to provide sufficient air to support combustion in the combustion section, or burners. Once ignition and fuel have been introduced and the light-off has occurred, the starter must continue to assist the engine until the engine reaches a self-sustaining speed. The torque supplied by the starter must be in excess of the torque required to overcome compressor inertia and the friction loads of the engine's compressor. Figure 10.1 illustrates a typical starting sequence for a gas turbine engine, regardless of the type of starter employed. As soon as the starter has accelerated the compressor sufficiently to establish airflow through the engine, the ignition is turned on followed by the fuel. The exact sequence of the starting procedure is important since there must be sufficient airflow through the engine to support combustion before the fuel-air mixture is ignited. At low engine cranking speeds, the fuel flow rate is not sufficient to enable the engine to accelerate; for this reason, the starter continues to crank the engine until self-accelerating speed has been attained. If assistance from the starter was cut off below the self-accelerating speed, the engine would either fail to accelerate to idle speed or might even decelerate, because it could not produce sufficient energy to sustain rotation or to accelerate during the initial phase of the starting cycle. The starter must continue to assist the engine considerably above the self-accelerating speed to avoid a delay in the starting cycle, which will result in a hot or hung false start or a combination of both. At the proper points in the sequence, the starter and ignition are automatically cut off. The basic types of starters that are in current use for gas turbine engines are direct current (DC) electric motors, starters/generators, and air turbine starters.

Common start method

Although the engine starting program is essentially the same, there are differences in the way it is implemented. Different power sources and starter types are also determined by the requirements of the engine and aircraft. In order to provide a stable

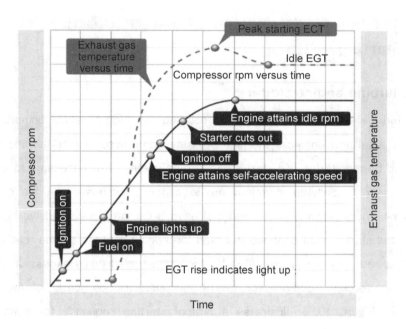

Figure 10.1 Typical starting sequence for a gas turbine engine

means of accelerating the engine rotor from its stationary state and supplying enough air and fuel for mixed combustion to the combustion chamber until the turbine can produce enough power to replace the starter's power, the starter must produce a high torque and pass it to the engine rotor. There are various kinds of starters:

Air turbine starter

At present, the air turbine starter is widely used in civil aviation engines. The air turbine starter is relatively light in weight and has a large torque, simple structure and use economy. The air turbine starter includes a single stage turbine, reducer, clutch and drive shaft, as shown in Figure 10.2. The power generated by the turbine is transmitted to the starter output shaft through the reducer and the clutch. The single stage turbine converts the air source pressure into mechanical energy. The reducer converts the high speed and low torque of the turbine output into the low speed and high torsion required for the engine rotor. The clutch controls the disconnection and engagement of the starter and the engine and prevents torque overload from damaging transmission components. The drive shaft transfers torque. The air turbine starter needs to have an air source. Its available air sources are the ground operated air cart, the engine auxiliary power unit (APU) and another running engine, as shown in Figure 10.3.

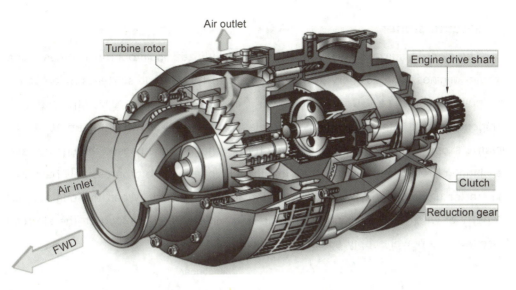

Figure 10.2　Cutaway view of an air turbine starter

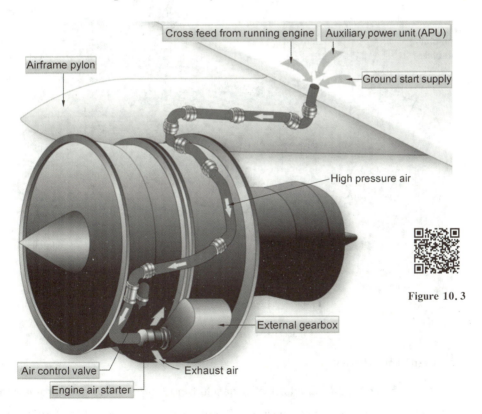

Figure 10.3

Figure 10.3　Air turbine starter are supplied by the ground operated air cart, APU, and another running engine

Electric starter

The electric starter is easy to use, mostly utilized in turboprop engines, small turbojet engines, and auxiliary power assemblies. The electric starter is made up of a DC power source, a DC electric motor, a speed reducer, a clutch, and other components. The internal structure of an electric starter is shown in Figure 10.4. The electric supply may be of a high voltage or low voltage and is passed through a system consisting of relays and resistors to allow the full voltage to be progressively built up as the starter gains speed. It also powers the ignition system. When the engine starts properly or when the beginning time cycle is completed and the load on the starter is lessened, the power supply automatically turns off.

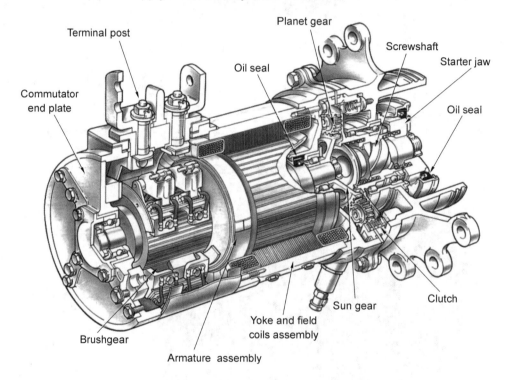

Figure 10.4 An electric starter

Gas turbine starter

Figure 10.5 shows the structure of a gas turbine starter. The gas turbine starter has a small and compact gas turbine. The engine typically has a turbine-driven centrifugal compressor, a reverse flow combustion chamber, and a mechanically independent free-power turbine. It has its own fuel and ignition system, starting method (often electric or hydraulic), and oil system. It is frequently quick, expensive, and difficult to control, and is typically utilized in some combats.

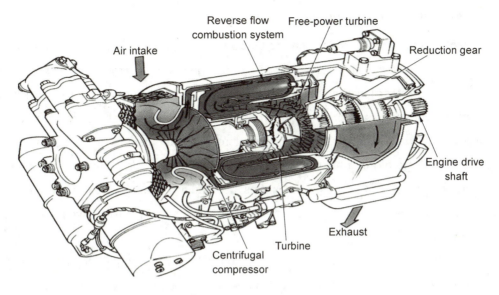

Figure 10.5 A gas turbine starter

Air impingement starting

Some turbo-jet engines do not have starter motors and instead use air impingement to rotate the turbine blades. Air from an external source or running engine is routed through non-return valves and nozzles onto turbine blades. Figure 10.6 shows a standard air impingement beginning method.

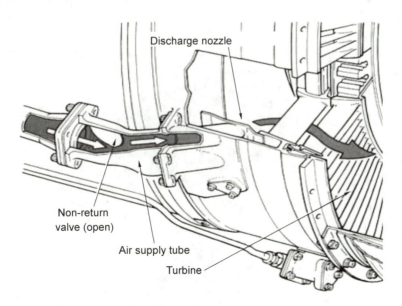

Figure 10.6 Air impingement starting

Generator starting system

Electric starting systems for gas turbine aircraft are of two general types: direct cranking electrical starting systems and starter generator starting systems. Direct cranking electric starting systems are used mostly on small turbine engines, such as auxiliary power units, and some small turboshaft engines. Many gas turbine aircraft are equipped with starter generator starting systems. Starter generator starting systems are also similar to direct cranking electrical starting systems, in addition to functioning as a starter, they contain a second series of windings that allow it to switch to a generator after the engine has reached a self-sustaining speed. This saves weight and space on the engine.

The starter generator is permanently engaged with the engine shaft through the necessary drive gears, while the direct cranking starter must employ some means of disengaging the starter from the shaft after the engine has started. The starter generator unit is basically a shunt generator with an additional heavy series winding (Figure 10.7). This series winding is electrically connected to produce a strong field and a resulting high torque for starting. Starter generator units are desirable from an economic standpoint, since one unit performs the functions of both starter and generator. Additionally, the total weight of starting system components is reduced and fewer spare parts are required.

Figure 10.7　Typical starter generator

The starter generator internal circuit has four field windings: a series field (C field), a shunt field, a compensating field, and an commutating winding (Figure 10.8). During starting, the C field, compensating field, and commutating windings are used. The unit is similar to a direct cranking starter since all of the windings used during starting are in series with the source. While acting as a starter, the unit makes no practical use of its shunt field. A source of 24 volts and 1,500 amperes is usually required for starting. When operating as a generator, the shunt field, compensating field, and commutating windings are used. The C field winding is used only for starting purposes. The shunt field winding is connected in the conventional voltage control circuit for the generator. Compensating and commutating or interpole windings provide almost sparkless commutation from no load to full load. Figure 10.9 illustrates the external circuit of a starter generator with an undercurrent controller. This unit controls the starter generator when it is used as a starter. Its purpose is to assure positive action of the starter and to keep it operating until the engine is rotating fast enough to sustain combustion. The control block of the undercurrent controller contains two relays. One is the motor relay that controls the input to the starter; the other, the undercurrent relay, controls the operation of the motor relay.

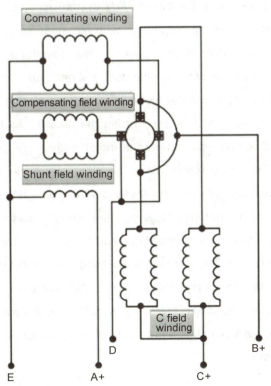

Figure 10.8　Starter generator internal circuit

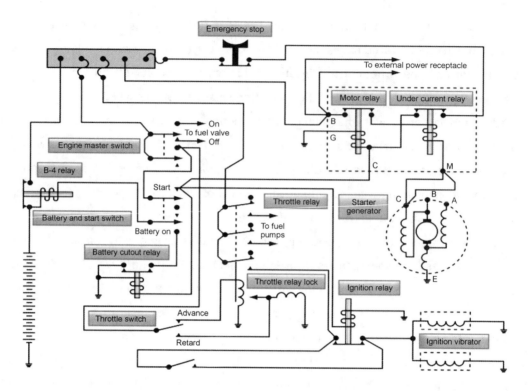

Figure 10.9 Starter generator circuit

The sequence of operation for the starting system is discussed in the following paragraphs. To start an engine equipped with an undercurrent relay, it is first necessary to close the engine master switch. This completes the circuit from the aircraft's bus to the start switch, to the fuel valves, and to the throttle relay. Energizing the throttle relay starts the fuel pumps and completing the fuel valve circuit provides the necessary fuel pressure for starting the engine. As the battery and start switch is turned on, three relays, namely the motor relay, ignition relay, and battery cutout relay close. The motor relay closes the circuit from the power source to the starter motor; the ignition relay closes the circuit to the ignition units; the battery cutout relay disconnects the battery. Opening the battery circuit is necessary because the heavy drain of the starter motor would damage the battery. Closing the motor relay allows a very high current to flow to the motor. Since this current flows through the coil of the undercurrent relay, the relay closes. Closing the undercurrent relay completes a circuit from the positive bus to the motor relay coil, ignition relay coil, and battery cutout relay coil. The start switch is allowed to return to its normal off position, and all units continue to operate.

10.2 Turbine ignition system

Turbine ignition systems are often less trouble-prone than reciprocating engine ignition systems since they are mostly used for a short while during the engine-starting cycle. It is not necessary to time the turbine engine ignition system to ignite at a specific moment in the operating cycle. The turbine engine ignition system is utilized to light the fuel in the combustor before being turned off. Under specific flight situations, other modes of operation for the turbine ignition system are employed, such as continuous ignition, which is performed at a lower voltage and energy level. Continuous ignition is used to prevent engine flameouts. This ignition can relight the fuel, preventing the engine from stopping. Critical flight modes that require continuous ignition include takeoff, landing, and some emergency operation modes. Gas turbine engines typically use a high-energy capacitor-type ignition mechanism and are cooled by fan airflow. Fan air enters the exciter box, passes around the igniter, and returns to the nacelle area. Cooling is necessary when using continuous ignition for a lengthy period of time. Gas turbine engines may use an electronic ignition system, a version of the simpler capacitor-type system. Turbine engines often use a capacitor-type ignition system, which consists of two independent components powered by a common low-voltage (DC) source such as the aircraft battery, 115 V AC supply, or permanent magnet generator. The engine immediately turns the generator via the accessory gear box, producing power at all times. The fuel in turbine engines can be easily ignited in ideal atmospheric circumstances, but because they frequently run in low temperatures at high altitudes, the system must be capable of supplying a high heat intensity spark. The ignition mechanism is highly reliable under diverse variables such as altitude, atmospheric pressure, temperature, fuel vaporization rate, and input voltage. This is achieved by supplying high voltage to the arc across a large spark gap.

A typical ignition system (Figure 10.10) consists of two ignition exciters, two intermediate ignition leads, and two high tension leads. To ensure safety, the ignition mechanism uses two igniter plugs.

Figure 10.11 depicts an older model capacitor-type turbine ignition system. The ignition exciters socket receives a 24 V DC supply. The electrical energy is filtered before reaching the exciter unit to prevent noise voltage from entering the aircraft's electrical system. Low-voltage input electricity powers a DC motor that drives one

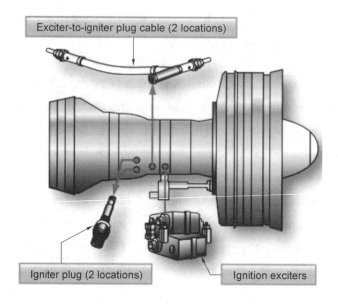

Figure 10.10　A typical ignition system

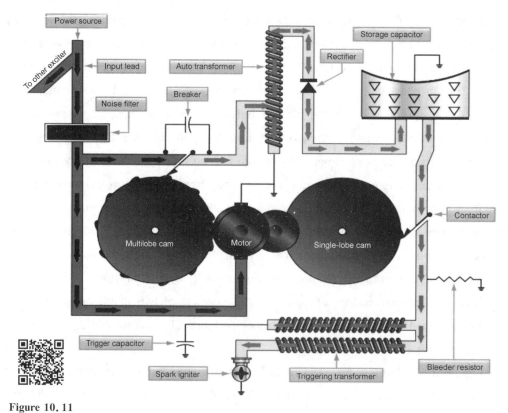

Figure 10.11　An older model capacitor-type turbine ignition system

multilobe cam and one single-lobe cam. The multilobe cam activates breaker points with input power. An auto transformer receives rapidly interrupted current from the breaker points. When the breaker closes, current flows through the primary winding of the transformer, creating a magnetic field. When the breaker opens, current stops flowing and the field collapses, causing voltage in the transformer's secondary. This voltage induces a pulse of current to flow into the storage capacitor via the rectifier, limiting the flow to one direction. With repeated pulses, the storage capacitor charges up to a maximum of about 4 joules (Note that one joule per second equals one watt.) The storage capacitor is connected to the spark igniter via the triggering transformer and a usually open contactor.

Once the capacitor is charged, the single-lobe cam closes the contactor mechanically. A portion of the charge passes via the primary of the triggering transformer and the capacitor attached to it, the resulting current generates a high voltage in the secondary, ionising the gap at the spark igniter.

When the spark igniter is made conductive, the storage capacitor releases its remaining energy and charge by connecting it in series with the triggering transformer's primary. The spark rate at the spark igniter fluctuates proportionally with the voltage of the DC power source, which influences the motor's rpm. However, because both cams are geared to the same shaft, the storage capacitor always absorbs energy from pulses before being discharged. Using a high-frequency triggering transformer with a low-reactance secondary winding minimizes discharge time. Concentrating maximum energy in minimal time creates an optimal spark for ignition, capable of destroying carbon deposits and vaporizing fuel globules.

The high voltage in the triggering circuits is totally segregated from the primary circuits. The exciter is hermetically sealed to protect components from harsh operating conditions and prevent flashover at altitude caused by pressure changes. This shielding prevents high-frequency voltage leakage that could interfere with airplane radio reception.

Capacitor discharge exciter unit

This capacity-type system provides ignition for turbine engines. Like other turbine ignition systems, it is required only for starting the engine; once combustion has begun, the flame is continuous. The energy is stored in capacitors. Each discharge circuit incorporates two storage capacitors, both are located in the exciter unit. The voltage across these capacitors is stepped up by transformer units. At the instant of igniter plug firing, the resistance of the gap is lowered sufficiently to permit the larger capacitor to discharge across the gap. The discharge of the second capacitor is of low-voltage, but of very high energy. The result is a spark oil of great heat intensity, capable of not only igniting abnormal fuel mixtures but also burning away any foreign deposits on the plug electrodes. The exciter is a dual unit that produces sparks at each of the two igniter plugs. A continuous series of sparks is produced until the engine starts. The power is then cut off, and the plugs do not fire while the engine is operating other than on continuous ignition for certain flight conditions. The exciters are air cooled to prevent overheating during long use of continuous ignition. Figure 10.12 shows a fan-air cooled exciter.

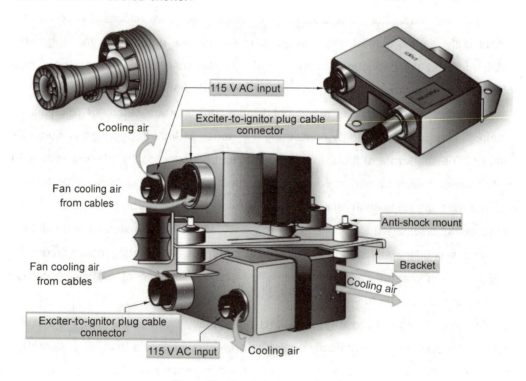

Figure 10.12 Fan-air cooled exciter

Igniter plugs

The igniter plug (Figure 10.13) of a turbine engine ignition system differs considerably from the spark plug of a reciprocating engine ignition system. Its electrode must be capable of withstanding a current of much higher energy than the electrode of a conventional spark plug. This high energy current can quickly cause electrode erosion, but because of the short operating time, igniter-related maintenance is reduced to a minimum. The electrode gap of the typical igniter plug is designed much larger than that of a spark plug since the operating pressures are much lower and the spark can arc more easily than in a spark plug. Furthermore, electrode fouling, common to the spark plug, is minimized by the heat of the high-intensity spark. Figure 10.14 shows a cutaway illustration of a typical annular-gap igniter plug, sometimes referred to as a long reach igniter because it projects slightly into the combustion chamber liner to produce a more effective spark. Another type of igniter plug, the constrained-gap plug, is used in some types of turbine engines. It operates at a much cooler temperature because it does not project into the combustion-chamber liner. This is possible because the spark does not remain close to the plug, but arcs beyond the face of the combustion chamber liner.

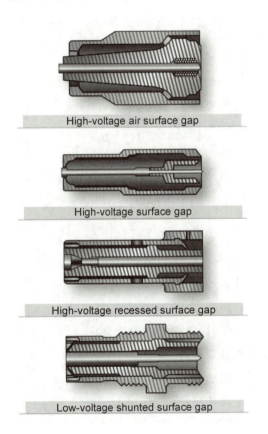

Figure 10.13 Ignitor plug

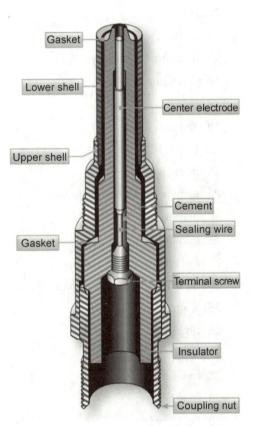

Figure 10.14 Typical annular-gap igniter plug

 Words & Expressions

1. start up　　　　　　　　　　启动
2. idling state　　　　　　　　慢车状态

课文翻译

3. accessory gearbox   附件齿轮箱
4. self-acceleration state   自加速状态
5. expelled   被喷射
6. dual spool   双转子
7. starter   启动机
8. as soon as   一旦
9. airflow   气流
10. DC motor   直流电动机
11. implement   实施
12. air turbine starter   空气涡轮启动机
13. reducer   减速器
14. clutch   离合器
15. mechanical energy   机械能
16. drive shaft   驱动轴
17. relay   继电器
18. lessen   缩小,减轻
19. turn off   关闭
20. compact   紧凑的
21. turbine driven centrifugal compressor   涡轮驱动离心式压气机
22. combat   战斗
23. air impingement starting   空气冲击式启动
24. direct cranking electrical system   直接启动式电气系统
25. winding   线圈
26. timing   定时(确定提前点火角)
27. relight   重新点火
28. exciter unit   激励器
29. igniter plug   火花塞,点火嘴
30. flash over   跳火
31. electrode erosion   电极腐蚀

Exercise

Ⅰ. Fill in the following blanks according to the text.

1. The starting mechanism is utilized to move the engine from a static to a steady

_____, with the starter carrying the engine rotor through the accessory gearbox.

2. When the speed achieves _____, the fuel system begins to provide oil, combining the air entering the combustion chamber with the fuel _____ from the nozzle to generate a fuel-air mixture.

3. On _____, axial flow engines, the high pressure compressor and N1 turbine system is only rotated by the _____.

4. The exact sequence of the starting procedure is important since there must be sufficient _____ through the engine to support combustion before the fuel-air mixture is ignited.

5. The basic types of starters that are in current use for gas turbine engines are _____, starter/ generators, and the air turbine type of starters.

Ⅱ. Translate the following sentences into Chinese.

1. Although the engine starting program is essentially the same, there are differences in the way it is implemented.

2. The air turbine starter includes a single stage turbine, reducer, clutch and drive shaft.

3. The electric supply may be of a high voltage or low voltage and is passed through a system consisting of relays and resistors to allow the full voltage to be progressively built up as the starter gains speed.

4. When the engine starts properly or when the beginning time cycle is completed and the load on the starter is lessened, the power supply automatically turns off.

5. The engine typically has a turbine-driven centrifugal compressor, a reverse flow combustion chamber, and a mechanically independent free-power turbine.

Lesson 11　Engine Fire-protection System

课文音频

Because fire is one of the most dangerous threats to an airplane, all current multi-engine aircraft contain a fixed fire protection system. A "fire zone" is a section or region of an aircraft that the manufacturer has identified as requiring fire detection and/or extinguishing equipment, as well as a high level of inherent fire resistance. The term "fixed" means that the fire protection system is a permanently installed system as opposed to any kind of movable fire extinguishing equipment, such as a hand-held extinguisher.

Several general problems or risks can cause overheating or flames in turbine engine aircraft due to their operational characteristics. There are two basic types of turbine failure: thermodynamic and mechanical.

Thermodynamic factors can alter the amount of air utilized to decrease combustion temperatures to levels that turbine materials can withstand. When the cooling cycle is disrupted, turbine blades may melt, resulting in a sudden loss of thrust. The quick accumulation of ice on inlet screens or inlet guide vanes can cause significant overheating, causing turbine blades to melt or get severed and hurled outward. Such failure can cause a severed tail cone and possible penetration of the aircraft structure, tanks, or equipment around the turbine wheel. In general, most thermodynamic failures are caused by ice, excessive air bleed or leakage, or incorrect controls which lead the compressor to stall or use excess fuel.

Mechanical failures, such as fractured or thrown blades, can also lead to overheat conditions or fires. Thrown blades can puncture the tail cone, creating an overheat condition. Failure of forward stages of multi-stage turbines usually is much more severe.

The whole engine fire protection system is separated into two parts: the fire detection system and the fire extinguishing system. The fire detection system monitors the fire temperature, overheating temperature, smoke concentration, and possible fire region of the engine and body, as well as high-pressure hot air leakage. When the monitoring data exceeds the warning value, a visual and auditory warning is delivered, indicating the particular location of the fire.

11.1 Engine fire detection system

A variety of fire detection systems are installed in aircraft to detect engine fires. The spot detection system and continuous-loop system are two often utilized types. Spot detection systems use individual sensors to monitor a fire area. Spot detection technologies include thermal switch fire detection, thermocouple fire detection, optical fire detection, and pneumatic-based thermal fire detection. Continuous-loop systems are often used in transport airplanes to provide comprehensive fire detection coverage with several loop-type sensors.

Thermal switch system

In a thermal switch system, the aircraft power system supplies one or more lights, and thermal switches control their operation. These thermal switches are heat-sensitive units that complete electrical circuits at specific temperatures. They are connected in parallel, but in series with the indicator lights (Figure 11.1). If the temperature in any segment of the circuit climbs over a predetermined level, the thermal switch closes, completing the light circuit and indicating a fire or overheat condition.

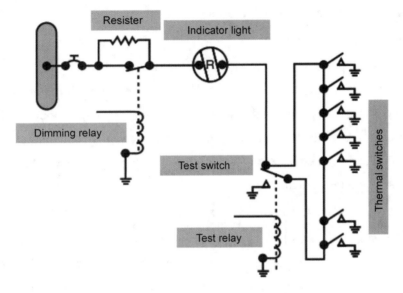

Figure 11.1 Thermal switch fire circuit

Thermocouple system

The thermocouple system acts differently than the thermal switch system. A thermocouple is dependent on the rate of temperature rise and does not warn when an engine gradually overheats or a short circuit occurs. The system includes a relay box, warning lights, and thermocouples.

Optical fire detection system

Optical sensors, often known as "flame detectors", detect certain radiation emissions from hydrocarbon flames and then generate an alarm. Optical sensors are classified into two types based on the emission wave lengths they are designed to detect: infrared (IR) and UV.

Infrared optical fire protection system

Infrared (IR) optical fire detectors are commonly employed in light turboprop aircraft and helicopter engines. These sensors are reliable and cheap to use for low-risk applications.

Pneumatic thermal fire detection system

Pneumatic detectors operate on the basis of gas laws. The sensing element is made up of a closed helium-filled tube with one end attached to a responder assembly. As the element is heated, the gas pressure inside the tube rises until the alert threshold is achieved. At this point, an internal switch closes, signaling an alarm to the cockpit. If the pneumatic detector loses pressure (for example, due to a leak), the integrity pressure switch opens and the fault alert sounds.

Continuous-loop system

Continuous-loop detectors can monitor fire alarms in huge regions, such as engines, and perform well. In practice, multiple continuous-loop fire detectors in different places are frequently linked together to cover a broader area of fire detection. This continuous fire detector connected in series is known as a temperature sensing loop. The detection principle of the temperature-sensing loop is identical to that of a single continuous-type detector.

Fenwal continuous-loop system

The Fenwal continuous-loop system comprises a thin inconel tube filled with thermally sensitive eutectic salt and a nickel wire center conductor. These sensing devices are connected in sequence to a control unit. They might be of equal or varied lengths and temperatures. The control unit, which operates directly from the power supply, applies a modest voltage to the sensor devices. When an overheat condition occurs anywhere along the element's length, the resistance of the eutectic salt within the sensing element reduces dramatically, allowing current to pass between the outer sheath and the core conductor. The control unit senses the current flow and generates a signal to activate the output relay. When the fire is extinguished or the critical temperature is reduced, the Fenwal continuous-loop system automatically switches to standby mode, ready to detect any following fire or overheat condition. The Fenwal continuous-loop system can be wired with a "loop" circuit. In this situation, if an open circuit occurs, the system will still indicate fire or overheating. If several open circuits develop, only the area between the breaks becomes inoperable.

Kidde continuous-loop system

The Kidde continuous-loop system involves two wires embedded in an inconel tube filled with a thermistor core material. The core has two electrical conductors that run its entire length. One conductor is grounded to the tube, while the other is connected to the fire detection control unit.

As the temperature of the core rises, the electrical resistance to the ground lowers. The fire detection control unit measures this resistance. If the resistance falls below the overheat set point, an overheat alert appears on the flight deck. Typically, a 10-second time delay is used to indicate overheating. If the resistance diminishes closer to the fire set point, a fire alert is issued. When the fire or overheat condition is removed, the resistance of the core material rises to the reset point, and the flight deck signs disappear. The rate of change in resistance indicates an electrical short or a fire. Resistance diminishes faster with an electrical short than with a fire. The Kidde continuous-loop system not only detects fires and overheats, but also sends nacelle temperature data to the aircraft in-flight monitoring system (AIMS).

11.2 Fire zone

The powerplant installation has several designated fire zones: (1) the engine power section; (2) the engine accessory section; (3) except for reciprocating engines, any complete powerplant compartment in which no isolation is provided between the engine power section and the engine accessory section; (4) any APU compartment; (5) any fuel-burning heater and other combustion equipments; (6) the compressor and accessory sections of turbine engines; and (7) combustor, turbine, and tailpipe sections of turbine engine installations that contain lines or components carrying flammable fluids or gases. Figure 11.2 shows fire protection for a large turbofan engine.

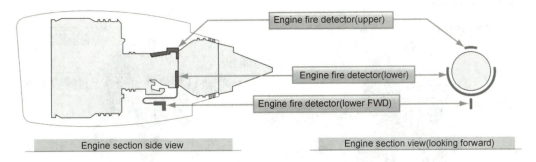

Figure 11.2　Large turbofan engine fire zone

11.3 Engine fire extinguishing system

The engine fire extinguishing system typically employs portable fire extinguishing bottles and fixed fire extinguishing bottles installed in specific areas to ensure safety and extinguish the fire as quickly as possible. Generally speaking, most aircraft have fixed fire extinguishing systems in the engine, APU, toilet, and cargo compartment, as well as portable fire extinguishing bottles in the cabin and cockpit.

Fire extinguishing agent

Most engine fire extinguishers employ inert agents to dilute the environment and prevent combustion. Many methods distribute the extinguishing ingredient through perforated tubing or discharge nozzles. High rate of discharge (HRD) systems employ open-ended tubes to release a large amount of extinguishing material in 1－2 seconds. Halon 1301 is the most commonly used extinguishing agent today due to its great firefighting capability and very low toxicity.

Containers

Fire extinguisher containers (HRD bottles) (Figure 11.3) are typically made of stainless steel and contain a liquid halogenated extinguishing chemical and pressurized gas (usually nitrogen). Alternative materials, such as titanium, can be used based on design requirements. Containers are available in various capacities and meet related criteria or exemptions. Most airplane containers are helical in shape, resulting in the least possible weight. However, cylindrical shapes are accessible if space is limited. Each container has a temperature/pressure sensitive safety relief diaphragm that prevents container pressure from exceeding the container test pressure in the event of high temperatures.

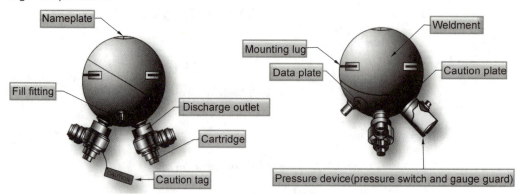

Figure 11.3　Fire extinguisher containers (HRD bottles)

Discharge valve

Discharge valves are mounted on the containers. The discharge valve assembly's outlet involves a cartridge (squib) and a frangible disk valve. Special assemblies with solenoid or manually actuated seat valves are also available. There are two procedures for releasing cartridge disks. The standard release type ruptures a segmented closure disk with an explosive-powered slug. For high temperature or hermetically sealed devices, a direct explosive impact cartridge ruptures a prestressed corrosion-resistant steel diaphragm through fragmentation. Most containers employ typical metallic gasket seals to permit refurbishing after discharge.

Two-way check valve

Two-way check valves are available in both lightweight aluminum and steel. These valves are needed in a two-shot system to keep the agent in a reserve container from backing up into the previously emptied main container.

Discharge indicator

Discharge indicators give visible proof of container discharge on fire extinguishing systems. There are two types of indicators: thermal and discharge. Both versions are intended for aircraft and skin mounting.

Thermal discharge indicator (red disk)

The thermal discharge indicator, connected to the fire container relief fitting, emits a red disk when the container's contents are thrown overboard due to high temperatures. When the disk blows out, the agent discharges through the opening created. This tells the flight and maintenance teams that the fire extinguisher container has to be renewed before the next trip.

Yellow disk discharge indicator

Activating the fire extinguisher system causes a yellow disk to be released from the airplane fuselage. This indicates to the maintenance team that the fire extinguishing system has been engaged by the flight crew, and that the fire extinguishing container must be replaced before the next flight.

Fire switch

Fire switches are normally located on the flight deck's center overhead panel or center panel. When the fire switch is activated, the engine stops because the fuel control is turned off, the engine is disconnected from the aircraft systems, and the fire extinguishing system is activated. Some aircraft employ fire switches that must be pulled and spun to activate the system, but others use a push-button switch with a guard. A lock is added to prevent the fire switch from being accidentally activated. The fire switch is only released when a fire is detected. The flight crew can manually release this lock if the fire detection system fails.

Warning system

To inform the flight crew, warning devices with visible and auditory signals are mounted in the cockpit. A horn sounds and one or more warning lights illuminate, informing the flight crew that an engine fire has been detected. These signals cease when the fire is extinguished.

 Words & Expressions

课文翻译

1. fire zone   防火区
2. hand-held extinguisher   手持式灭火器
3. thermodynamic   热能动力学
4. melt   融化
5. overheating   过热
6. hurl outward   向外甩出
7. puncture   刺穿
8. fire detection system   探火系统
9. fire extinguishing system   灭火系统
10. continuous-loop system   连续型系统
11. spot detection system   点式探测系统
12. thermal switch   热开关
13. thermocouple   热电偶
14. optical fire detection   光学火焰探测器
15. pneumatic-based thermal fire detection 气动式热火焰探测器
16. relay box   继电器盒
17. IR (infrared)   红外线
18. UV (ultraviolet)   紫外线
19. sensing element   感应元件
20. closed helium-filled tube   封闭的充满氦气的管子
21. inconel tube   铬镍铁管子

22. modest voltage　　　　　　　合适的电压
23. thermistor　　　　　　　　　热敏材料
24. flight deck　　　　　　　　　飞机驾驶舱
25. AIMS（aircraft integrated　　飞机飞行监控系统
    monitoring system）
26. dilute　　　　　　　　　　　稀释
27. pop-out　　　　　　　　　　弹出式
28. discharge valve　　　　　　　排放阀
29. cartridge　　　　　　　　　　爆炸帽
30. frangible　　　　　　　　　　法兰盘；易碎的
31. blow out　　　　　　　　　　吹飞

 Exercise

Ⅰ. Fill in the following blanks according to the text.

1. A "_____" is a section or region of an aircraft that the manufacturer has identified as requiring fire detection and/or extinguishing equipment, as well as a high level of inherent fire resistance.

2. The term "fixed" means that the fire protection system is a permanently installed system as opposed to any kind of movable fire extinguishing equipment, such as a _____.

3. The quick accumulation of ice on inlet screens or inlet guide vanes can cause significant overheating, causing turbine blades to melt or get severed and _____ outward.

4. The whole engine fire protection system is separated into two parts: the _____ and the _____.

5. Spot detection technologies include _____ fire detection, _____ fire detection, _____ fire detection, and pneumatic-based thermal fire detection.

Ⅱ. Translate the following sentences into Chinese.

1. The Fenwal continuous-loop system comprises a thin inconel tube filled with thermally sensitive eutectic salt and a nickel wire center conductor.

2. Containers are available in various capacities and meet related criteria or exemptions.

3. The discharge valve assembly's outlet involves a cartridge (squib) and a frangible disk valve.

4. When the disk blows out, the agent discharges through the opening created.

5. Activating the fire extinguisher system causes a yellow disk to be released from the airplane fuselage.

Lesson 12　Engine Indication System and Ice Protection System

课文音频

Engine instruments measure the operating parameters of aero engines. These are often quantity, pressure, speed, and temperature indicators. The most popular engine instruments are fuel and oil pressure gauges, tachometers (percent calibrated), and temperature gauges (such as exhaust gas temperature, turbine inlet temperature gauges). In addition, engine instruments found on turbine-powered aircraft include engine pressure ratio gauges, fuel quality indicators, fuel flow meters, N1 and N2 compressor speed indicators, torquemeters (on turboprop and turboshaft Engines).

The engine instrumentation is generally located in the center of the flight deck for easy visibility by the pilot and copilot (Figure 12.1). This may not be true for light aircraft with only one flying crewmember. Multi-engine aircraft frequently employ a single gauge for a specific engine characteristic, they display information for all engines using numerous pointers on the same dial face.

Figure 12.1　Engine instrumention

12.1 Engine instrument

Engine oil pressure gauge

The engine oil pressure gauge(Figure 12.2) is the most important instrument used by pilots to check the health of their engines. Oil pressure is often measured in pounds per square inch (psi, 1 psi = 6.89 kPa). The normal operating range typically is represented by a green arc on the circular gauge. The exact permitted operating range should be determined by reference to the manufacturer's operating and maintenance data. In reciprocating and turbine engines, oil is used to lubricate and cool bearing surfaces where parts rotate or slide past each other at high speeds. A loss of pressurized oil in these regions would quickly result in excessive friction and overheating, leading to catastrophic engine failures.

Figure 12.2　Oil pressure gauge

Engine pressure ratio gauge

Turbine engines include pressure indicators that correlate with power output. This is known as the engine pressure ratio (EPR) gauge(EPR indicator). The EPR gauge is a differential pressure gauge that measures total exhaust pressure to ram air pressure at the engine's inlet, calculates engine thrust based on temperature, altitude, and other factors, and then compares these two pressures. The pressure ratio transmitter in the EPR gauge uses a bellows arrangement to compare these two pressures and transmit an electric signal for the gauge to indicate.

Fuel pressure gauge

Fuel pressure gauges provide important information to pilots (Figure12.3). Fuel is often pumped from the aircraft's numerous tanks to power the engines. If a fuel pump malfunctions or a tank is drained beyond its capacity to sustain output pressure, the pilot must take immediate action. While direct-sensing fuel pressure gauges using Bourdon tubes, diaphragms, and bellows sensing devices are available, it is particularly undesirable to insert fuel lines into the flight deck due to the potential risk of fire from leaks. For optimal results, the sensing mechanism should be integrated into a transmitter device that sends a signal to the flight deck indicator via electricity. In some cases, fuel flow rate indicators are utilized instead of fuel pressure gauges.

Figure 12.3 Fuel pressure gauge

Turbine gas temperature indicating system

Exhaust gas temperature (EGT) is an essential parameter for turbine engine functioning. The EGT signaling system on the flight deck displays the temperature of turbine exhaust gases as they exit the unit. In some turbine engines, the EGT is recorded near the unit's entrance. In this case, the EGT signaling system is also known as the turbine inlet temperature (TIT) signaling system. The EGT or TIT is measured using several thermocouples (Figure 12.4). They are placed at regular intervals around the engine turbine casing or exhaust duct. The tiny thermocouple voltages are often amplified and utilized to power a servomotor, which drives the indication pointer. It is customary to base a digital drum indicator on the motion of the pointer. The EGT indicator depicted in Figure 12.4 is hermetically sealed. The instrument's scale goes from 0 °C to 1,200 °C, and there is a vernier dial in the top right corner, as well as a power off warning flag in the lower portion.

Fuel quantity indicating system

All aircraft fuel systems must provide a fuel quantity indication. The complexity of

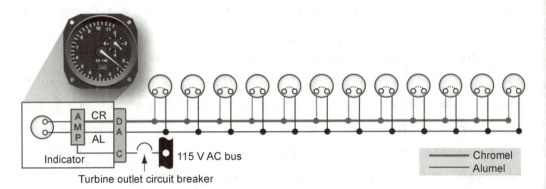

Figure 12.4 A typical exhaust gas temperature signaling system

a fuel system and the aircraft on which it is mounted greatly impact the devices installed. The first quantity indicators were simple and did not require electricity. They are still widely used today. Direct reading indicators are only suitable for light aircraft with fuel tanks near to the flight deck. Electric or electronic capacitance-type indicators are required for larger and other light aircraft.

Fuel flow meter

A fuel flow meter monitors an engine's fuel usage in real time. Pilots can use this information to evaluate engine performance and plan flights. The type of fuel flow meter used on an airplane depends on the motor and fuel system. Turbine-engine aircraft have the largest range in fuel density due to the temperature and composition of the fuel. These aircraft employ an elaborate fuel flow mechanism. It measures gasoline mass to provide correct fuel flow indication during flight.

Tachometer

The tachometer measures the speed of the crankshaft in a reciprocating engine. A calibrated dial on a direct or remote-indicating instrument indicates revolutions per minute (rpm). The tachometer in the reciprocating engine monitors engine power and ensures them operates within certified limits. Gas turbine engines also have tachometers. They monitor the speed of the engine's compressor section(s). Turbine engine tachometers are calibrated in percentage of rpm, with 100% equal to the optimal turbine speed. This provides for consistent operating methods across various engine rpms. In addition to the engine tachometer, helicopters use a tachometer to indicate main rotor shaft rpm. It should also be noted that many reciprocating-engine tachometers also have built-in numeric drums that are geared to the rotational mechanism inside. These are hour meters that keep track of the time the engine is operated. There are two types of tachometer system in wide use today: mechanical and electrical.

Mechanical Tachometers: Mechanical tachometer indicating devices are used in small, single-engine light aircraft with a short distance between the engine and the instrument panel. They consist of an indicator connected to the engine via a flexible driving shaft. The driving shaft is geared into the engine so that it turns when the engine does. The indicator has a flyweight assembly connected to a gear mechanism that powers a pointer. As the driving shaft rotates, centrifugal force works on the flyweights, causing them to move to an angled position. This position fluctuates with the engine's rpm. The gear system transmits the amount of movement of the flyweights to the pointer. The rotating pointer on the tachometer indicates the engine's rpm.

Electric tachometers : A mechanical coupling between the engine and rpm indicator is not feasible for aircraft with engines not located in the fuselage, just forward of the instrument panel. Electric tachometers provide greater accuracy and require less maintenance. There are several different electric tachometer systems that can be used, so it is necessary to consult the manufacturer's instructions for further information on each one. A common electric tachometer system utilizes a tiny AC generator attached to the gear case of a reciprocating engine or the accessory drive section of a turbine engine. The generator turns in tandem with the engine. The generator's frequency output is directly proportional to engine speed. It is wired to a synchronous motor in the indicator, which replicates the output. A drag cup or drag disc link is used to drive the indicator as in a mechanical tachometer. Gas turbine engines also have tachometers. They monitor the speed of the engine's compressor section(s).

12.2　Ice protection system

Ice protection systems are designed to operate using one of two possible modes, either in anti-icing mode, where ice formation is prevented, or in de-icing mode, where ice is removed once it has formed. An ice protection system can prevent the deformation of ice in three distinct ways, namely thermally, mechanically or chemically, so as to limit the amount of ice to an acceptable range. Thermal and chemical ice prevention systems can be designed to operate in either anti-icing or de-icing modes. However, mechanical ice protection systems work in the de-icing mode once ice has formed, but before the formations have reached dangerous levels. For anti-icing to be maintained, the thermal or chemical system must be operated continuously. Should such a system be operated intermittently, then it would provide de-icing protection.

Engine bleed air thermal ice protection

Engine bleed air is the most common source of ice protection on medium and large sized transport aircraft. The turbofan engines powering such aircraft can supply the required quantities of hot, high pressure air to provide ice protection. Figure 12.5 shows a schematic diagram for a bleed air thermal ice protection system for a large transport aircraft wing. This shows hot air being directed to the leading edge of the wing through a pipe which has holes in it to allow hot air to flow onto the inner surface of the wing's leading edge. The air is typically pressurised at approximately 20 psi. Control is achieved by electrically operated valves, and temperature monitoring is incorporated to provide system feedback to prevent dangerous overheating. It can be seen from Figure 12.5 that only the outboard part of the wing is protected. As a rule of thumb, chords above five meters do not require protection, as long as the wing has a reasonably large sweep. Similarly the lip of each engine intake would be protected on such an aircraft.

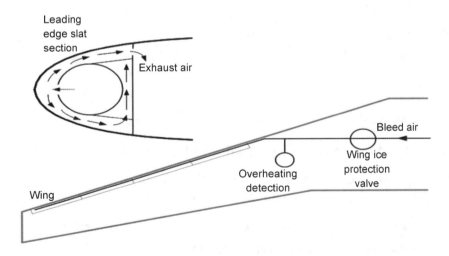

Figure 12.5 Wing anti-ice control

Exhaust heated ram air thermal ice protection

Exhaust heated ram air thermal ice protection system relies upon a suitable source of exhaust heat. Turboprop engines with long exhausts are particularly suitable to this application, while turbofan engine, with typically short exhausts are not. The modern high bypass ratio turbofan engines in use today can only emply this system if a heat exchanger is installed in the short exhaust system, and installing a heat exchanger means a significant loss of engine performance. Further disadvantages of this type of ice protection system include the fact that forward speed is needed for operation. Additionally, air pressures and quantities are limited, and the system does not easily accommodate engine intake protection.

Words & Expressions

课文翻译

1. EGT   排气温度
2. TIT   涡轮进口温度
3. tachometer   转速表
4. torquemeter   扭矩表
5. EPR gauge   发动机压力表
6. crankshaft   曲轴
7. mechanical tachometer   机械式转速表
8. flyweight   飞重（机械调速器中用于调节转速的部件）
9. tandem   串联（指两个或多个设备或系统按顺序连接）
10. anti-icing mode   防冰模式
11. de-icing mode   除冰模式
12. chord   翼弦线（机翼前缘到后缘的直线距离）

Exercise

Ⅰ. Fill in the following blanks according to the text.

1. The most popular engine instruments are fuel and _____, _____, and _____.

2. A loss of pressurized oil in these regions would quickly result in _____ and _____, leading to catastrophic engine failures.

3. Turbine engines include pressure indicators that correlate with power output. This is known as the _____ (EPR) gauge (EPR indicator).

4. The tiny _____ are often amplified and utilized to power a servomotor, which drives the indication pointer.

5. The _____ on the flight deck displays the temperature of turbine exhaust

gases as they exit the unit.

6. An ice protection system can prevent the deformation of ice in three distinct ways, namely _____, _____ or _____, so as to limit the amount of ice to an acceptable range.

7. _____ is the most common source of ice protection on medium and large sized transport aircraft.

8. _____ relies upon a suitable source of exhaust heat.

9. The _____ in use today can only employ this system if a heat exchanger is installed in the short exhaust system, and installing a heat exchanger means a significant loss of engine performance.

10. Control is achieved by _____, and temperature monitoring is incorporated to provide system feedback to prevent dangerous overheating.

II. Translate the following sentences into Chinese.

1. Engine instruments measure operating parameters of aero engines.

2. The engine instrumentation is generally located in the center of the flight deck for easy visibility by the pilot and copilot.

3. Multi-engine aircraft frequently employ a single gauge for a specific engine characteristic, they display information for all engines using numerous pointers on the same dial face.

4. The normal operating range typically is represented by a green arc on the circular gauge.

5. While direct-sensing fuel pressure gauges using Bourdon tubes, diaphragms, and bellows sensing devices are available, it is particularly undesirable to insert fuel lines into the flight deck due to the potential risk of fire from leaks.

6. Ice protection systems are designed to operate using one of two possible modes, either in anti-icing mode, where ice formation is prevented, or in de-icing mode, where ice is removed once it has formed.

7. Thermal and chemical ice prevention systems can be designed to operate in either anti-icing or de-icing modes.

8. However, mechanical ice protection systems work in the de-icing mode once ice has formed, but before the formations have reached dangerous levels.

9. Turboprop engine, with long exhausts are particularly suitable to this application, while turbofan engine, with typically short exhausts are not.

10. Additionally, air pressures and quantities are limited, and the system does not easily accommodate engine intake protection.

Glossary of Commonly-used Terms

常用专业
词汇音频

abradable	耐磨的
absorption	吸收
accessory	附件
accessory drive	附件驱动器
accessory gearbox	附件齿轮箱
afterburning	二次燃烧,后燃
AIMS (aircraft integrated monitoring system)	飞机飞行监控系统
air impingement starting	空气冲击式启动
air intake duct	进气道
air turbine starter	空气涡轮启动机
airflow	气流
airfoil	翼型
airport	飞机场
airstream	气流
alloy	合金
ambient	周围的,周围环境的
annular	环状的
anticorrosion	防腐
anti-icing mode	防冰模式
as soon as	一旦
atomization	雾化
attachment	连接
attenuate	(使)变小,减弱
axial	轴向的
bacteria	细菌
bearing	轴承
blast	爆炸
blisk	整体叶盘
blockage	堵塞

blow out	吹飞
bolt	螺栓,用螺栓固定
breather subsystem	通气的子系统
bulk	大部件
bulkhead	舱壁,隔板,隔框
casing	框,壳
calibrate	校准
can-annular-type burner	环管型火焰筒
capacity	容量
cartridge	爆炸帽
centrifugal	离心的
centrifugal force	离心力
chord	翼弦线(机翼前缘到后缘的直线距离)
circumferential	圆周的
clamp	夹具
clearance	间隙
clog	阻塞,堵塞
closed helium-filled tube	封闭的充满氦气的管子
cloud point	浊点,云点
clutch	离合器
coaxially	同轴地
combat	战斗
combustion	燃烧
compact	紧凑的
consecutive	连续的,不间断的
consume	消耗
continuously loop system	连续型系统
convergent	收缩的
corrosion	腐蚀,侵蚀
corrugated	波纹的
cowl	整流罩
crack	破裂,变形
crankshaft	曲轴

creep	蠕变
creep plate	蠕盘
cross-sectional area	横截面面积
cycle	使……循环
DC motor	直流电动机
de-aerated	去除蒸汽的
debris	残骸，碎片
deflection	偏移
deformation	变形
de-icing mode	除冰模式
dilute	稀释
dilution	稀释物
direct cranking electrical system	直接启动式电气系统
disc	圆盘状物
discharge	排除
discharge valve	排放阀
dissipate	驱散
disproportionately	不相称，不成比例地
distribution	分布
divergent	发散的
drainage	排水，放水；排水系统
drive cam	驱动凸轮
drive shaft	驱动轴
dry sump variable pressure lubrication system	干式油底壳可变压力润滑系统
dual spool	双转子
eccentric bore	偏心孔
EGT	排气温度
electrode erosion	电极腐蚀
elevation	提高，升高
encompass	包含，包括
endoscope	内窥镜
EPR indicator	发动机压力表
exciter unit	激励器

exhaust	消耗，耗尽
exhaust nozzle	喷管
exhaust sleeve	排气套管
exhaust system	排气系统
expelled	被喷射
expose	揭发，使暴露
exterior	外部的
external	外部的
extract	获得，提取
FADEC(full authority digital electronic control)	全权限数字电子控制
fan cowling	风扇整流罩
fastener	紧固件
fatigue	疲劳
favoured	受优惠的，有特权的
fence	围栏
fine mesh	细滤网
fire detection system	探火系统
fire extinguishing system	灭火系统
fire zone	火区
flame blowout	熄火
flame cartridge	火焰筒
flame out	熄火
flange	法兰
flare	喇叭形口
flash over	跳火
flash point	闪点
flexibility	灵活性
flight deck	飞机驾驶舱
flow divider	分流器
flyweight	飞重（机械调速器中用于调节转速的部件）
forge	锻造
frangible	法兰盘；易碎的

freezing point	冰点
frontal area	最大截面
fuel heater	燃油加热器
fuselage	（飞机的）机身
gasoline	汽油
gear pump	齿轮泵
gerotor pump	摆线泵
gradient	梯度
hand-held extinguisher	手持式灭火器
helicopter	直升机
high bypass ratio	高涵道比（指涡扇发动机中旁通流与核心流的比例较高）
hold-down nut	压紧螺母
hollow	孔洞的，中空的
honeycomb tube	蜂窝管
housing	腔，室
hurl outward	向外甩出
hush	安静
hybrid	混合的
hydromechanical fuel control	液压机械式燃油控制
hydromechanical/electronic fuel control	液压机械/电子式燃油控制
idling state	慢车状态
igniter	点火器，电嘴
igniter plug	火花塞，点火嘴
impede	妨碍，阻碍
impel	推动
impeller	叶轮
implement	实施
impulse	冲击
in order to	为了，目的在于
inconel tube	铬镍铁合金管
ingress	进入
inhalation	（空气等的）吸入
injection	注入，喷射

insulate	隔绝，隔热
IR（infrared）	红外线
kerosine	航空燃油
kinetic	运动的，运动引起的
leakage	泄露
lend	借
lessened	缩小的，减轻的
linger	徘徊
lobe	叶轮
lubrication performance	润滑性能
magnetic chip detector	磁屑探测器
Mach number	马赫数
manifold	集气管，多支管
manufacture	制造
mechanical	机械的
mechanical energy	机械能
mechanical tachometer	机械式转速表
melt	融化
military	军事的
minimal	最小的
mitigate	减轻
mixer	混合器
modest voltage	合适的电压
moisture	湿润
molecule	分子
nacelle	飞机的引擎机舱
negate	使无效，否定
non-supercharged	非增压的
noxious	有害的，有毒的
nozzle	喷嘴
nozzle diaphragm	喷管挡栅板
oblique	（线或角）斜的，倾斜的
occupant	占有人，居住者，乘客
oil cooler	滑油散热器

oil tank	滑油箱
on the lip of	在……的边缘
operator	运营商
optical fire detection	光学火焰探测器
optimum	最适宜的,最佳的
orifice	孔,洞口
overheating	过热
perforate	穿孔于,在……上打眼
perimeter	周长;周围,边界
periphery	边缘,外围
permanent	永久的
petroleum oil	从石油中提炼的滑油
pipe	管道
plenum chamber	增压室
pneumatic	气动的
pneumatic-based thermal fire detection	气动式热火焰探测器
pod	(飞机的)吊舱
pollution	污染
pop-out	弹出式
pressurization	增压
propagation	传播,传输
propeller(＝propellor)	螺旋桨,推进器
psi(pound per square inch)	(压强单位)磅每平方英寸
puncture	刺穿
purge	清除,(使)净化
radiator	散热器
ram	撞击,冲
rectifying cone	整流锥
reducer	减速器
regulator	监管机构
relay	继电器
relay box	继电器盒
relight	重新点火
remainder	剩余物

retrofit	翻新,改型
reverser	换向器
right angle	直角
rivet	铆钉
robust	结实的,坚固的
rotodynamic pump	回转动力式泵
roughness	粗糙度
runway	飞机跑道
scavenge	清除污物
secondary manifold	二级歧管
segment	分段
self-acceleration state	自加速状态
self-aligning	自动对中(准)
sensing element	感应元件
serration	锯齿状,锯齿状突起
shroud	保护罩
shroudless	无罩的
snout	锥形进口
spatial	空间的
spindle	轴
spot detection system	点式探测系统
spray	喷洒
spring	弹簧
stack	大量
stagnant	不流动的,停滞的
stainer	油滤
stall	(飞机的)失速
start up	启动
starter	启动机
stealth	隐形的
stiffen	(使)变硬,(使)强硬
Strato power pump	斯特拉托动力泵
streamline	把……做成流线型
subsonic	亚声速的

suction	吸,吸出
sufficient	足够的
supercharge	对……增压,涡轮增压
suppression	抑制
suppressor	干扰抑制器
surpass	超过
swash plate	旋转斜盘,倾斜盘
swirl vanes	旋流器
synthetic oil	人工合成滑油
tailpipe	排气管
tachometer	转速表
tandem	串联(指两个或多个设备或系统按顺序连接)
tank	燃料箱
tapered	锥形的
thanks to	由于
thermal	热的
thermal switch	热开关
thermistor	热敏材料
thermocouple	热偶
thermodynamic	热能动力学
throat	咽喉,颈前部
thrust	推力
thrust reverser	推力反向器
timing	定时(确定提前点火角)
tiny crystals	微小的晶体
TIT	涡轮进口温度
toroidal	环形的
torquemeter	扭矩表
trim	修整,完善;配平
turbine	涡轮
turbine driven centrifugal compressor	涡轮驱动离心式压气机
turbo supercharged	装有涡轮增压器的
turbulence	湍流

turn off	关闭
unevenness	不均匀
UV (ultraviolet)	紫外线
vane	叶片
vane pump	叶片泵
ventilation	通风,通气
volatility	挥发性
volume	体积
vortex	涡流
water absorption performance	吸水性能
weld	焊接
wet-sump lubrication system	湿式油底壳滑油系统
winding	线圈

课后习题答案

Lesson 1　Classifications of Aero Engines

Ⅰ.1. propellers　2. powerplant　3. Compression　4. diffuser　5. bypass ratio

Ⅱ.1. 在燃煤/燃气炉或核反应堆中释放的能量被转移到水中,转化为蒸汽,蒸汽膨胀并驱动活塞或涡轮机。

2. 往复式发动机是一种机械装置,它在一个封闭的容器中混合燃料和空气,通过热气燃烧释放热能,然后热气膨胀,在连杆上做功。

3. 往复式发动机的主要部件是曲轴箱、气缸、活塞、连杆、气门、气门操作机构和曲轴。每个气缸盖中都有气门和火花塞。

4. 现代航空往复式发动机属于四冲程发动机。四冲程往复式发动机在四个冲程内完成四个过程:进气、压缩、做功和排气。

5. 目前,航空燃气涡轮发动机有四种基本类型:涡轮喷气发动机、涡扇发动机、涡轮螺旋桨发动机和涡轴发动机。

6. 大多数涡轴发动机通过多级自由涡轮驱动其输出轴,这种涡轮能够最大限度地从废气中提取能量。

Lesson 2　The Development of Aero Engines

Ⅰ.1. piston　2. spins　3. metallurgy　4. reliability　5. consumption

Ⅱ.1. 人类历史上第一台飞机发动机是1903年莱特兄弟发明的12马力四缸汽油发动机,它使用铝铸缸体,解决了在保持动力的同时减轻重量的难题。

2. 与传统活塞发动机相比,喷气发动机提供了更高的速度、更高的效率和更好的飞行性能。

3. 发动机被设计用于在高温高压下运行,同时提供卓越的推重比和燃油效率。

4. 电力推进、燃料电池和替代燃料等技术为未来的飞机发动机带来了希望,而专家们在寻求进一步突破航空的界限的同时,将继续解决安全和效率问题。

5. 该发动机采用了宽弦风扇、浮壁式燃烧室、单晶涡轮叶片和粉末冶金涡轮盘、叶尖间隙主动控制系统和全权限数字电子控制系统等先进设备,提高了发动机的竞争力。

Lesson 3　Air Inlet Duct

Ⅰ.1. thermal　2. fuselage　3. military　4. subsonic　5. spring

Ⅱ.1. 冷端的主要部件包括发动机的进气道和压气机,热端的主要部件包括燃烧室、涡轮、喷管和一些反推器。

2. 因此,在飞行中,每当飞行马赫数超过这个阈值时,进气道还起使气流减速的关键作用,有效地将气流的动能转变为压力的升高。

3. 随着飞机前进速度继续增加,冲压效应变得更大,即使由于进入发动机的空气速度增加而产生一些能量损失,推力也会增加。

4. 但是,当进气压力下降到低于环境空气压力的程度时,压差将迫使它们打开,并向压气机入口提供额外的空气。

5. 当空气被吸入进气导管时,通常会形成一个高能涡流,该涡流会产生强大的吸力,从机翼下方吊舱内的发动机处到达地面。

6. 进气道中形成一个或多个斜激波,使空气减速到接近声速,然后形成一个正常的激波,完成从超声速到亚声速的过渡。

Lesson 4　Compressors

Ⅰ.1. kinetic　2. impeller　3. manifold　4. alloy　5. annular　6. airfoil

Ⅱ.1. 大部分空气从压气机流入燃烧部分,其中一些被称为压气机引气的空气用于入口管道的防冰和热段部件的冷却。其他引气用于客舱增压、空气调节、燃油系统防冰和气动发动机启动。

2. 气流沿离开叶轮中心方向流动的叫作离心式压气机;气流沿与叶轮轴平行的方向流动的叫作轴流式压气机。

3. 离心式压气机具有坚固耐用、重量轻、易于制造以及各压缩级压力比高的特点,因此被广泛应用于许多早期的燃气涡轮发动机。

4. 无论使用什么术语,这些出口管道都在扩散过程中起着非常重要的作用,也就是说,它们将气流方向从径向改变为轴向,气体扩散过程在转弯后完成。

5. 叶片通常在前级进行分段组装,并可能在其内端添加内衬,以最大限度地减少气流变化造成的较长叶片的振动。

Lesson 5　Combustion Section

Ⅰ.1. combustion　2. coaxially　3. flare　4. propagation　5. atomization

Ⅱ.1.其他类型发动机在发动机后部也有部分高速射流排出,但大部分推力或功率是由附加的涡轮级驱动大风扇、螺旋桨或直升机旋翼产生的。

2.由于在正常混合比下航空燃油燃烧的速度仅为每秒几英尺,即使在空气扩散流中被点燃的燃油,都会以每秒80英尺的速度被吹走。

3.这种低速再循环采用环形涡流的形式,类似于烟圈,具有稳定和锚定火焰的作用。

4.主气流(约占气流总质量的四分之一)进入燃烧过程,流入内衬,在那里其速度被降低到不足以吹灭燃料燃烧的火焰的程度。

5.但它们的缺点是涡轮机温度不均匀,如果发生故障,所产生的极端温差可能会导致涡轮机故障。

Lesson 6　Turbine Section

Ⅰ.1. accessories　2. deflections　3. aerofoil　4. hollow　5. abradable

Ⅱ.1.涡轮吸收能量的确切数值取决于其驱动的负载情况,具体包括压气机的尺寸和类型、附件数量以及其他涡轮所施加的载荷。

2.涡轮盘中的应力随着速度的平方而增加,因此为了在较高的速度下保持相同的应力水平,必须不成比例地增加截面厚度,从而增加重量。因此,最终设计是在效率和重量之间折中的结果。

3.喷管导叶和涡轮叶片通道的设计广泛基于空气动力学考虑,为了获得最佳效率,与压气机和燃烧设计兼容,喷管导叶和涡轮叶片具有基本的翼型形状。

4.涡轮叶片采用翼型设计,旨在在相邻叶片之间提供通道,使气流能够稳定加速至"喉部"区域。该区域面积最小,气流速度达到出口所需,从而产生所需的反作用力。

5.用于涡轮系统冷却的空气从压气机较高的一级排出,虽然其温度高于1000°F,但它比驱动涡轮机的气体冷得多。这些空气流过空心叶片,并与废气一起排出。

Lesson 7　Exhaust Section

Ⅰ.1. suppressor　2. orifice　3. scavenging　4. nacelle　5. corrugated　6. cowl

Ⅱ.1.在涡轮喷气发动机中,排气依靠其速度和压力产生推力,但在涡轮螺旋桨发动机中,只有少量的推力是由废气产生的,因为大部分气流的能量已被吸收,以推动涡轮螺旋桨旋转。

2.为了确保恒定重量(或体积)的气体能够以声速流过任何给定点,超声速排气管的后部被扩大,以容纳以超声速流动的气体的额外重量或体积。否则,喷管将无法高效运行。

3. 尽管集热器系统提高了排气系统的背压,但涡轮增压带来的马力增益远远抵消了背压增加导致的马力损失。短排气管系统相对简单,其拆卸和安装主要包括拆卸和安装压紧螺母和夹具。

4. 飞机和发动机制造商与监管机构及运营商携手合作,不断改进每一代产品的降噪技术。在条件允许的情况下,还会将新技术应用于早期机型的改造中,以降低噪声。

5. 在这些发动机上,进气道产生的声音通常会比尾管产生的声音更大。此外,涡扇发动机的进气道和排气管内部都衬有消声材料,用以降低噪声水平。

Lesson 8　Fuel System

Ⅰ. 1. de-aerated, gasoline　2. Freezing point　3. flash point　4. combustion　5. moisture, pollution, tank, bacteria

Ⅱ. 1. 燃油压力系统、燃油控制系统和燃油指示器系统是组成发动机燃油系统的三个功能齐全的子系统。

2. 如果燃油的冰点不够低,很容易导致燃油内部产生冰晶,从而导致燃油流量降低,影响发动机的正常供油。

3. 闪点太低,燃油的稳定性差,飞机容易着火。

4. 如果油箱内残留水就会滋生细菌,从而导致油箱腐蚀。

5. 燃烧室内的燃油压力可以提升到燃油足以被高压系统有效雾化的水平。

Lesson 9　Lubrication and Cooling Systems

Ⅰ. 1. roughness　2. anticorrosion　3. synthetic oil, petroleum oil　4. bearing　5. capacity

Ⅱ. 1. 虽然每个部分的接触面看起来都很光滑,但在显微镜下仍显现出一定的粗糙度。

2. 滑油系统的主要用途是防腐、清洁、冷却和润滑。

3. 发动机滑油有两种类型,即人工合成油和石油基滑油,它们常用于活塞发动机。

4. 压力系统向主引擎轴承和附件驱动器提供滑油。

5. 燃气涡轮发动机通常会有一个单独的油箱连接到发动机外壳的一个容易接近的区域,允许路线维护人员完成他们的(维护)工作。

Lesson 10　Starting and Ignition Systems

Ⅰ.1.idling state　2.self-acceleration state，expelled　3.dual-spool,starter　4.airflow　5.direct current electric motors

Ⅱ.1.尽管发动机的启动程序本质上是相同的,但在实施方式上有所不同。

2.空气涡轮启动机包括单级涡轮、减速器、离合器和传动轴。

3.通过由继电器和电阻组成的系统,电动式启动机可以由高压或低压电力驱动,可以在(启动机)启动加速积累的时间段内传递全部电压到启动机。

4.当发动机正常启动或启动时间周期结束,启动机负载减轻时,电源自动关闭。

5.发动机通常有一个涡轮驱动离心式压气机、一个回流燃烧室和一个机械独立的自由动力涡轮。

Lesson 11　Engine Fire-protection System

Ⅰ.1.fire zone　2.hand-held extinguisher　3.hurled　4.fire detection system，fire extinguishing system　5.thermal switch，thermocouple，optical

Ⅱ.1.Fenwal连续型系统包括一个充满热敏共熔盐的镍铬铁合金管和一个镍丝中心导体。

2.防火瓶有多种容量可供选择,并需要符合相关标准或豁免规定。

3.排放阀组件的出口包括一个爆炸帽和一个易碎的蝶阀。

4.当安全阀中的圆盘(因压力过高而)破裂时,灭火剂会通过因破裂而产生的开口快速释放。

5.激活灭火系统会使一个黄色圆盘从机身上被释放出来。

Lesson 12　Engine Indication System and Ice Protection System

Ⅰ.1.oil pressure gauges，tachometers，temperature gauges　2.excessive friction,overheating　3.engine pressure ratio　4.thermocouple voltages　5.EGT signaling system　6.thermally，mechanically，chemically　7.Engine bleed air　8.Exhaust heated ram air thermal ice protection system　9.modern high bypass turbofan engines　10.electrically operated valves

Ⅱ.1.发动机仪表测量飞机发动机的运行参数。

2.发动机仪表盘通常位于驾驶舱中间,以便飞行员和副驾驶看到。

3.多引擎飞机经常采用单一仪表来衡量特定的发动机特性,它们在同一个仪表盘上使用多个指针显示所有发动机的信息。

4.正常工作范围通常用圆形仪表上的绿色圆弧表示。

5.虽然可以使用波登管、空气膜盒等直接传感装置,但由于泄漏可能引起火灾风险,因此人们特别不希望将燃油管路引入驾驶舱。

6.防冰保护系统有两种可能的工作模式,一种是防冰模式(防止冰形成),另一种是除冰模式(冰一旦形成就会被去除)。

7.热引气防冰和化学防冰系统可以设计成在防冰或除冰模式下运行。

8.然而,机械防冰保护系统是在冰形成后且结冰量达到危险水平之前进行除冰。

9.长排气管道涡桨发动机特别适合这种应用,而具有典型短尾排气管道的涡扇发动机则不行。

10.此外,气压和流量是有限的,因此该系统并不容易适应发动机进气道保护。

参考文献

[1] 蔡景.航空发动机构造与维修管理[M].北京:北京航空航天大学出版社,2015.

[2] 黄燕晓,瞿春红.航空发动机原理与结构[M].北京:航空工业出版社,2015.

[3] 陈光.航空发动机发展综述[J].航空制造技术,2000(6):24-27,34.

[4] 金伟.世界航空发动机发展趋势及经验[J].中国工业评论,2016(11):38-44.

[5] 张津,洪杰,陈光.现代航空发动机技术与发展[M].北京:北京航空航天大学出版社,2006.

[6] 许春生.燃气涡轮发动机[M].北京:兵器工业出版社,2006.

[7] 西北工业大学,南京航空学院,北京航空学院.航空燃气涡轮发动机原理[M].北京:国防工业出版社,1981.

[8] LEFEBVRE A H, BALLAL D R. Gas turbine combustion: Alternative fuels and emissions[M]. 3rd ed. New York: CRC Press, 2010.

[9] CRANE D. Aviation maintenance technician series: Powerplant[M]. 3rd ed. Washington: Aviation supplies & Academics, Inc, 2013.

[10] US department of Transportation, Federd Auiation Administration. Aviation maintenance technician handbook — Powerplant. Volume 2[M].[S. l.: s. n.]. 2014.

[11] WILKINSON R. Aircraft structure and systems[M]. London: Prentice Hall, 1996.

[12] KOROLEV V V, LISKOVSKAYA E V, PAVLOV D A. Calculation and simulation of a starter-generator for the aviation gas turbine engine[C]//2020 XXIII International Conference on Soft Computing and Measurements (SCM). Piscataway: IEEE, 2020:249-251.

[13] ROYCE R. The jet engine[M]. 5th ed. Hoboken: Wiley-Black Well, 2015.

[14] SFORZA P M. Theory of aerospace propulsion[M]. Oxford: Butterworth-Heinemann, 2011.